T0214015

SpringerBriefs in Materials

The SpringerBriefs Series in Materials presents highly relevant, concise monographs on a wide range of topics covering fundamental advances and new applications in the field. Areas of interest include topical information on innovative, structural and functional materials and composites as well as fundamental principles, physical properties, materials theory and design. SpringerBriefs present succinct summaries of cutting-edge research and practical applications across a wide spectrum of fields. Featuring compact volumes of 50 to 125 pages, the series covers a range of content from professional to academic. Typical topics might include

- A timely report of state-of-the art analytical techniques
- A bridge between new research results, as published in journal articles, and a contextual literature review
- A snapshot of a hot or emerging topic
- An in-depth case study or clinical example
- A presentation of core concepts that students must understand in order to make independent contributions

Briefs are characterized by fast, global electronic dissemination, standard publishing contracts, standardized manuscript preparation and formatting guidelines, and expedited production schedules.

More information about this series at http://www.springer.com/series/10111

Lobna A. Elseify · Mohamad Midani ·
Ayman El-Badawy · Mohammad Jawaid

Manufacturing Automotive Components from Sustainable Natural Fiber Composites

 Springer

Lobna A. Elseify [iD]
Faculty of Engineering and Materials
Science
German University in Cairo
New Cairo, Egypt

Ayman El-Badawy [iD]
Faculty of Engineering and Materials
Science
German University in Cairo
New Cairo, Egypt

Mohamad Midani [iD]
Faculty of Engineering and Materials
Science
German University in Cairo
New Cairo, Egypt

Wilson College of Textiles
NC State University
Raleigh, USA

Mohammad Jawaid [iD]
Biocomposite Technology Laboratory
Institute of Tropical Forestry and Forest
Products (INTROP)
Universiti Putra Malaysia
Selangor, Malaysia

ISSN 2192-1091   ISSN 2192-1105   (electronic)
SpringerBriefs in Materials
ISBN 978-3-030-83024-3   ISBN 978-3-030-83025-0   (eBook)
https://doi.org/10.1007/978-3-030-83025-0

This Springer imprint is published by the registered company Springer Nature Switzerland AG
The registered company address is: Gewerbestrasse 11, 6330 Cham, Switzerland

To my family, my number one supporters. Thank you for all the love and support you have always given me.

Lobna A. Elseify

Foreword

Over 80 million passenger vehicles, which are mostly powered by petroleum-based products, were produced in 2018 worldwide according to the Organisation Internationale des Constructeurs d'Automobiles (OICA), and the number is expected to significantly increase to meet the predicted increase in demand. This poses serious challenge in terms of greenhouse emissions. It is estimated that road transport is responsible for nearly 30% of the EU's total CO_2 emissions. Moreover, millions of these vehicles reach the end of their use each year, and in September 2000, the End of Life Vehicles Directive was officially adopted by the EU Parliament. Putting more pressure on the automobile manufacturers to reduce fuel consumption and to consider the end-of-life options of their vehicles. To mitigate the negative impact of the use of petroleum-based products, electrical vehicles operated with batteries as well as super capacitors have been developed. However, improvement is sorely needed for such cars for traveling longer distance before recharging. One key parameter in this vein is to reduce the overall weight of the automotive components by using high-performance fiber-reinforced composites from natural renewable materials that are inexpensive, biodegradable, and environmentally friendly.

Natural fiber composites (NFC) provide the solution needed since they are not only sustainable, but they are also lightweight and possess high strength to weight ratio. The research in NFC is challenging due to surface properties and non-uniformity in length, crimp, and linear density within and between natural fibers. Recently, numerous researchers contributed to the public domain literature. However, the gap between the academic literature and the actual practices in the industry is limiting the contribution of academic researchers in this area. Hence, it is important to have a compilation of the state-of-the-art knowledge on the use of NFC in the automotive industry.

Results reported in this monograph are part of the investigation striving to shed light on this topic. It includes compilation, tabulation, comparison, and analytical interpretation of the data from industry and academic literature. It includes the major areas of interest to academic experienced and young researchers, such as historical context of using NFC in the automotive industry, specific types of materials used and their qualification programs, major technical challenges facing NFC, major processing techniques and their parameters, analysis of major NFC parts used and their performance requirements, sustainability assessments including life cycle assessment and carbon footprint, in addition to future trends.

I endorse the work of the authors and have the pleasure to recommend this book to the readers.

2021   Prof. Abdel-Fattah M. Seyam
                                              TATM Department Head, Raleigh, NC, USA

Preface

The new race in the automotive industry is no longer speed, but rather lightweighting and sustainability. Major automakers and component suppliers are now shifting to natural fiber reinforcements for their composite components as a sustainable and lightweight alternative to the manmade reinforcements. However, there is a gap between academic literature and the actual practices in the industry. Therefore, the main objective of this book is to bridge this gap and to provide a comprehensive and integrated review on recent practices on natural fiber composites (NFC) in the automotive industry from composites fabrication to recycling.

Manufacturing Automotive Components from Sustainable Natural Fiber Composites book acts as a body-of-knowledge or handbook for researchers and industrialists interested in this field. It provides a historic context of the adoption of NFC in the automotive industry and provides technical data on the materials used, with up-to-date statistics on global production. Moreover, it discusses the process of materials qualification between material suppliers and automakers and the major challenges facing the use of NFC in the industry. Further, it provides a critical review on the manufacturing techniques used, including processing parameters and procedures. Afterward, it combines and tabulates the major manufactured NFC auto parts, with details on major performance requirements. The book also evaluates the sustainability of NFC, by comparing the life cycle assessments and carbon footprints of different components made with different natural fibers. Finally, it provides future prospects and trends in NFC and the use of all-cellulose composites.

This book can help both researchers and practitioners by removing the hindrance of having to search several studies. It provides academic researchers with a better understanding of recent practices of NFC in the automotive industry and help them spot the research gaps on which they will build their future research studies. Moreover, it is intended to help practitioners determine how to use NFC in producing new sustainable and innovative auto parts.

We are highly thankful to Springer Nature team for their generous cooperation at every stage of the book preparation and production.

New Cairo, Egypt   Lobna A. Elseify
New Cairo, Egypt   Mohamad Midani
New Cairo, Egypt   Ayman El-Badawy
Selangor, Malaysia   Mohammad Jawaid

Contents

Chapter 1
Natural Fibers in the Automotive Industry

Abstract The need to find a sustainable alternative to the man-made fibers used in reinforcing composites has led the automotive manufacturers and parts suppliers to shift to natural fiber reinforcements. Early versions of natural fiber composite (NFC) were only limited to non-structural parts; however, with the ongoing intensive research, the idea to use NFC in more structural and exterior parts is rapidly expanding. Vegetable fibers are the major class of natural fibers used in the automotive, due to their high specific properties in addition to their high thermal and acoustical insulation. This chapter is dedicated to discussing recent practices on natural fiber composites in the automotive industry. The history of adoption of NFC, from Henry Ford's hemp car in the 1930 till Polestar flax interior panels in 2020, is presented. The technical data on the materials used, including natural fibers and bio-based matrices, is discussed. Finally, up-to-date statistics on world production and global market of vegetable fibers and bioplastics is provided.

Keywords Natural fiber reinforcements · Bioplastics · Automotive · Composites

1.1 Introduction

Vegetable fiber, one of the three types of natural fibers, has three main constituents: cellulose, hemicellulose, and lignin. Vegetable fiber, which is the main scope of this book, could be subcategorizing based on the cellulosic fiber source or form as shown in Fig. 1.1. Cellulosic fibers could be extracted from the bast, leaf, fruit, or seed of the plant. They have two forms: native cellulose, referred to as cellulose I, and regenerated cellulose, referred to as cellulose II (Elseify et al. 2019). Differences between vegetable fibers, mainly depend on their chemical composition and their physical and mechanical properties. Bast fibers are better used as reinforcement in composites, since bast fibers' main function is to provide plants with support. As for leaf fibers, they have better toughness. Fruit or seed fibers provide composites with elastic toughness (Suddell 2008; Stokke et al. 2013; Elseify et al. 2019). Figure 1.2 shows some examples of common vegetable fibers used in automotive composites.

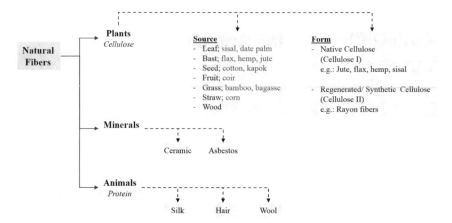

Fig. 1.1 Classification of natural fibers (Suddell 2008; Stokke et al. 2013; Elseify et al. 2019)

Fig. 1.2 Examples of common vegetable fibers used in automotive composites

The need to find a sustainable replacement to the man-made fibers used in reinforcing composites has drove the world to start using sustainable natural fibers. Primarily, there are two main issues with man-made fiber-reinforced composites: large quantities of CO_2 emissions, and non-biodegradability. However, natural fibers can offer solutions for these problems. First, they are biodegradable. Second, their CO_2 emission is very small compared to man-made fibers. Also, natural fiber composites (NFCs) are lighter than synthetic composites which will in return reduce the CO_2 emissions. Hence, natural fiber composites are regarded as greener alternative materials to replace already existing composites in different market sectors. Natural fiber composites are mostly used in the automotive industry and insulation. Figure 1.3

Fig. 1.3 Natural fiber composite revenue by application 2016 (US$742 mil)

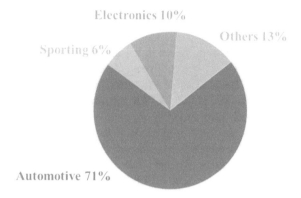

Electronics 10%

Others 13%

Sporting 6%

Automotive 71%

shows revenue by application in 2016 where 742 million US dollars were spent on natural fiber composites in various applications (Grand View Research 2018).

The leading automotive manufacturers and parts suppliers have been investigating the potential of using natural fibers as a reinforcement to synthetic or bio-based matrices since the start of the twentieth century. However, the use of natural fibers did not break through until the end of the twentieth century when the fossil fuels started to deplete. Unfortunately, the properties of natural fibers are lower than man-made fibers, namely glass, aramid, and carbon fibers. Consequently, it was a greater challenge to substitute all the synthetic composites in the automotive vehicles with natural fiber. Hence, natural fiber-reinforced composites were only limited to interior parts at first and they were only used in the manufacturing of non-structural parts like seat fillers, seat backs, spare tire cover, wood trims, headliners, dashboards, loudspeaker housings, parcel shelves, and thermal and acoustic insulations (Koronis et al. 2013; Carruthers and Quarshie 2014; Akampumuza et al. 2017). However, with the ongoing intensive research and the well-establishment of automotive interior parts, the idea to use natural fiber-reinforced composites in more structural and exterior parts is rapidly expanding. Natural fiber-reinforced composites could be used in the manufacturing of seat frames, load floors, floor pans, driver train, and steering components as will be discussed in this review (Suddell and Evans 2005; Akampumuza et al. 2017).

However, the review and research articles available in the academic literature only show the work done without linking the research to the industry, thus creating a gap between researchers and what actually happens in the industry. Therefore, the main objective of this book is to bridge this gap and to provide a comprehensive and practical review on recent practices on natural fiber composites in the automotive industry from composite fabrication till recycling.

The world is now shifting to the use of biodegradable natural materials in the automotive industry. This shift is for environmental and performance enhancement reasons. This section summarizes the history of using natural fiber composites in the automotive industry. Moreover, the thermoplastic or thermoset matrices and natural fibers used in the automotive industry will be discussed.

1.2   History

The use of natural fibers in the automotive industry dates back to the 1930s. The very first NFC in automotive applications was implemented by Henry Ford between 1930 and 1940s when he introduced the world's first car made from hemp-reinforced composites (Suddell 2008; Akampumuza et al. 2017; Witayakran et al. 2017; Mann et al. 2018; Bcomp 2020a, b, c, d, e; McLaren 2020; Porsche_Motorsport 2020). However, the mass production of such cars was not economical back then; hence, they switched back to manufacturing cars from metals. However, in the 1950s there was another attempt made by the East German Trabant Car, to use natural materials in cars, and it continued till 1990. Later, in the 1990s, Daimler Chrysler made several attempts to use natural fibers in the automotive industry. During the 1990s, coir fibers and latex were used together in the manufacturing of back seats, head restraints, bunk cushions, and sun visors. Afterward, in 1994, they used flax and sisal fibers in the interior trim components. Then, in 1996, jute fibers were used in the manufacturing of Mercedes-Benz E-class door panels. Not to mention that the use of natural fibers did not break through until the end of the twentieth century when the fossil fuels started to deplete. Since then, nearly all European reputed car manufacturers have taken the rout of using eco-friendly biodegradable recyclable materials in their car parts. Figure 1.4 shows a timeline of the history of natural fiber usage in the automotive industry (Suddell and Evans 2005; Suddell 2008; Akampumuza et al. 2017; Mann et al. 2018; Bcomp 2020c; McLaren 2020).

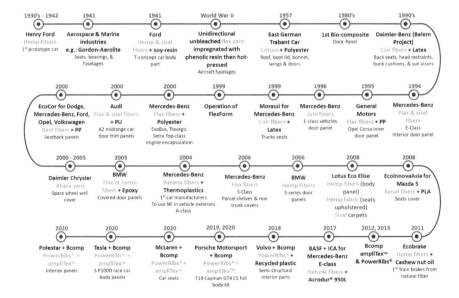

Fig. 1.4   History of using natural fibers in automotive industry

Till the start of the twenty-first century, natural fiber composites were only utilized in interior car parts. However, in 2004, Mercedes-Benz manufactured exterior parts made from banana fibers and thermoplastics, making A-class vehicles the 1st cars with natural fibers in exterior applications. Then, in 2012 and 2013, a company named Bcomp launched ampliTex™ and powerRibs®, and flax reinforcing preforms to be used in automotive applications. Afterward, Bcomp partnered with several car manufacturers like Volvo and McLaren in the manufacturing of semi-structural interior parts and car seats, respectively (Bcomp 2020c; McLaren 2020).

1.3 Materials

1.3.1 Natural Fiber Reinforcements

Natural fibers can be obtained from both animal and plant sources. The fibers obtained from plant sources are also known as vegetable fibers, and they are the most commonly used fiber reinforcements in automotive composites. In 2019, according to the FAO statistics, the world production of natural vegetable fibers excluding cotton was 6.786 million tonnes (FAO 2021). Figure 1.5a shows the distribution of vegetable fiber production by fiber type in 2019. Unfortunately, there is a lack of biodiversity in fiber crops, and more than 80% of vegetable fibers are obtained from only 3 crops: jute, flax, and coir. Similarly, 80% of vegetable fibers are obtained from only 4 countries: Bangladesh, India, France, and Vietnam as shown in Fig. 1.6, which illustrates the geographic distribution of vegetable fiber production in 2019. This lack of biodiversity has resulted in a slow growth rate in the world production of vegetable fibers over the past decades as shown in Fig. 1.7.

The most commonly used natural fibers in applications like automotive industry are flax, kenaf, hemp, jute, sisal, coir, and abaca fibers. In 2012, 30,000 tonnes of natural fibers was used in the automotive industry distributed as shown in Fig. 1.5b.

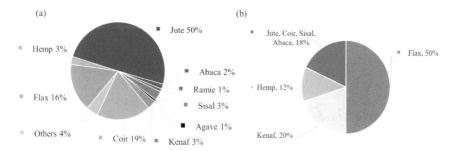

Fig. 1.5 **a** Distribution of world production of vegetable fibers excluding cotton by fiber type in 2019 (total quantity 6.786 million tonnes) (FAO 2021) and **b** utilization of natural fiber reinforcements in the automotive composites in 2012 (total volume of 30,000 tonnes) (de Beus et al. 2019)

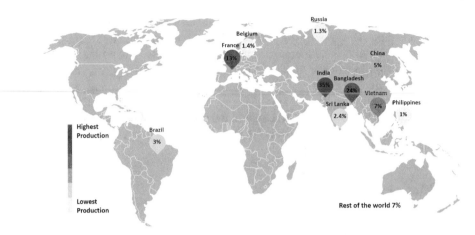

Fig. 1.6 Distribution of world production of vegetable fibers excluding cotton by country in 2019 (total quantity 6.786 million tonne)

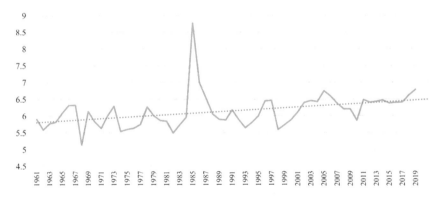

Fig. 1.7 World production of vegetable fibers excluding cotton from 1961 to 2019 obtained from FAO database

Flax fibers were used the most with 50% market share corresponding to nearly 15,000 tonnes. Kenaf fibers represented more than 20% of the market share, followed by hemp fibers then jute, coir, abaca, and sisal fibers (de Beus et al. 2019).

1.3.2 Matrix Material

Both thermoplastic and thermoset matrices are being used with natural fibers in automotive composites. Natural fibers have low thermal stability, which in turn limits the number of used thermoplastic polymers to three or four. In general, the polymers used must have processing temperature below 230 °C like polyolefins, polyethylene, polypropylene, and ethylene propylene rubber. The most commonly used thermoplastic matrix in the automotive industry is polypropylene. Figure 1.8 shows a classification of some polymeric matrix materials. On the other hand, epoxy resin was frequently used with natural fibers as thermoset matrix. Polyamides, polyester, and polycarbonates are not used as matrices for natural fibers because they require more than 250 °C during processing (Magurno 1999; Bledzki et al. 2002; Suddell and Evans 2005). The matrix is one of the factors that affects the properties of the produced composite. For instance, thermoplastics will help in producing recyclable composites unlike thermosets. Thermoplastics are of high interest in the automotive industry due to their short production cycle time; they do not require extra curing time as in case of thermosets. However, the processing of thermosets is much easier than thermoplastic. Moreover, one of the main problems with natural fibers is the poor interfacial bond between fibers and the matrix. The interfacial bond could be improved by treating the fiber surface and by adding some coupling agents (Pickering et al. 2016). The literature is full of research contributions to develop reliable composites to be used in industries like the automotive industry. However, there is a gap between the NFC currently used in the automotive industry and what researchers are investigating.

However, to take full advantage of NFC as bio-based, biodegradable, and sustainable, bioplastics could be used as matrices. Bioplastics, according to the association of European Bioplastics (Berlin), are plastics that are bio-based, biodegradable, or both. There are three groups of bioplastics: bio-based and non-biodegradable, bio-based and biodegradable, and biodegradable fossil-based plastics as shown in Fig. 1.9a. The advantage of using bioplastics over conventional plastics is that their carbon footprint is lower which will reduce the overall carbon footprint and greenhouse gas emissions of the final composites. Moreover, bioplastics have improved strength, breathability, and optical properties (European Bioplastics 2019). In 2019, the annual production of bioplastics was nearly equal to 2.114 million tonnes, 55.5% of which are biodegradable polymers and 44.5% are bio-based non-biodegradable

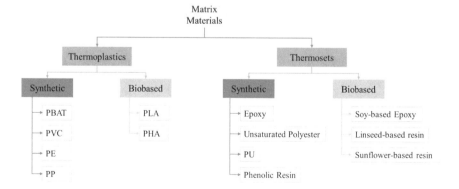

Fig. 1.8 Classification of polymeric matrix materials

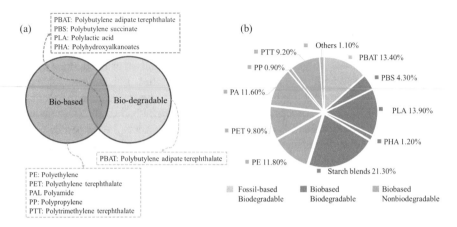

Fig. 1.9 a Bioplastic classifications and **b** global production capacity of bioplastics in 2019 (total volume of 2.11 million tonnes) (European Bioplastics 2019)

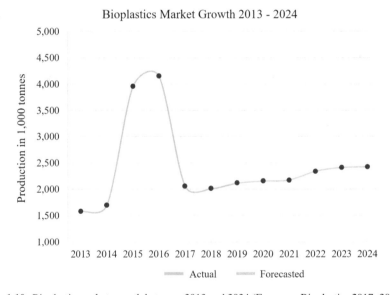

Fig. 1.10 Bioplastic market growth between 2013 and 2024 (European Bioplastics 2017, 2020)

polymers. The world production of bioplastics by material in 2019 is presented in Fig. 1.9b.

The European Bioplastics has published several interesting market reports and could be accessed through: https://www.european-bioplastics.org/market/. According to the data provided, the world production of bioplastics was 2.114 million tonnes in 2019; it is expected to reach 2.426 million tonnes in 2024. However, after analyzing the published data, it was found that the global production capacity has decreased from 4.165 million tonnes in 2016 to become 2.114 million tonnes in 2019 as shown in Fig. 1.10. The data in Fig. 1.10 was compiled from the reports of European Bioplastics and Nova Institute; it shows the annual production of bioplastics till 2019 and the forecasted numbers till 2024 (European Bioplastics 2017, 2020, 2021).

Recently, an innovative binder was developed by BSAF, a German company located in New Jersey, called Acrodur® specially for the NFC in the automotive industry. Acrodur® is a water-based formaldehyde-free binder. It is applicable for processes used with thermosets and thermoplastics. Acrodur® can be used with hemp, kenaf, flax, and wood. The binder is characterized by being eco-friendly. Moreover, the prepregs can be made using various techniques enabling fiber content up to 75%. It is used in the manufacturing of roof frames (interior application) and engine encapsulations (exterior application). In 2016, the 1st sun roof frame made from natural fibers (70 wt.%) and Acrodur® was invented by International Automotive Components (IAC) for Mercedes-Benz E-class (IAC Group 2020). Fifty percentage of the frame's weight was reduced when used instead of the metal frames. Another successful example is using Acrodur® with natural fibers in combustion engines. It has superior thermomechanical stability, and when added to natural fibers, noise absorption was improved. Acrodur® can be used in car seats, roof frames, door trims, and car carpets.

Another company named Sicomin, located in France, is also considered one of the leading companies in the production of a thermoset bio-epoxy resin called Sicomin GreenPoxy. They created several bio-based epoxy systems that have different properties depending on the intended applications. One version named SR FireGreen 37 is a fire-retardant epoxy system that is made from 25% carbon from plant and vegetable origin and has lower environmental impact than standard epoxy systems. Additionally, they have two other products with higher bio-based content: SR GreenPoxy 33 and SR GreenPoxy 56 which are composed of 35 and 50% bio-based components, respectively. Their products could be used in various manufacturing techniques such as resin transfer molding and hand layup, as well as in the manufacturing of prepregs.

References

Akampumuza O, Wambua PM, Ahmed A et al (2017) Review of the applications of biocomposites in the automotive industry. Polym Compos 38:2553–2569. https://doi.org/10.1002/pc.23847

Bcomp (2020a) ampliTex™. http://www.bcomp.ch/en/products/amplitex

Bcomp (2020b) PowerRibs™. http://www.bcomp.ch/en/products/powerribs. Accessed 18 May 2020

Bcomp (2020c) PowerRibs™ helps Volvo cars use ocean plastic in automotive interior parts

Bcomp (2020d) Racing the EGT Tesla and building a sustainable racetrack. https://www.bcomp.ch/news/racing-egt-tesla-and-building-a-sustainable-racetrack/

Bcomp (2020e) Porsche Motorsport innovates with series production of bio-based composites. https://www.bcomp.ch/news/porsche-innovates-with-series-production-of-bio-based-composites/. Accessed 3 Feb 2021

Bledzki AK, Sperber VE, Faruk O (2002) Natural and wood fibre reinforcement in polymers. Macromolecules 13:3–144

Carruthers J, Quarshie R (2014) Technology overview biocomposites. Knowl Transf Netw 70

de Beus N, Carus M, Barth M (2019) Carbon footprint and sustainability of different natural fibre for biocomposites and insulation material. Nov Inst 57

Elseify LA, Midani M, Shihata LA, El-Mously H (2019) Review on cellulosic fibers extracted from date palms (Phoenix Dactylifera L.) and their applications. Cellulose 26. https://doi.org/10.1007/s10570-019-02259-6

EuropeanBioplastics (2019) Bioplastics facts and figures

European Bioplastics (2017) Bioplastic market data 2017. Eur Bioplastics 1–7

European Bioplastics (2020) Bioplastics market data 2019

European Bioplastics (2021) Bioplastic materials. https://www.european-bioplastics.org/. Accessed 20 Feb 2021

FAO (2021) FAOSTAT. http://www.fao.org/faostat/en/#data/QC. Accessed 19 Feb 2021

Grand View Research (2018) Natural fiber composites (NFC) market size, share & trends analysis report by raw material, by matrix, by technology, by application, and segment forecasts, 2018–2024

IAC Group (2020) IAC FiberFrame™ natural fiber sun roof frame debuts on 2017 Mercedes-Benz E-Class. https://www.iacgroup.com/media/2016/04/04/iac-fiberframe-natural-fiber-sun-roof-frame-debuts-on-2017-mercedes-benz-e-class/. Accessed 23 Feb 2021

Koronis G, Silva A, Fontul M (2013) Green composites: a review of adequate materials for automotive applications. Compos Part B Eng 44:120–127. https://doi.org/10.1016/j.compositesb.2012.07.004

Magurno A (1999) Vegetable fibres in automotive interior components. Angew Makromol Chemie 272:99–107. https://doi.org/10.1002/(SICI)1522-9505(19991201)272:1%3c99::AID-APMC99%3e3.0.CO;2-C

Mann GS, Singh LP, Kumar P, Singh S (2018) Green composites: A review of processing technologies and recent applications. J Thermoplast Compos Mater. https://doi.org/10.1177/0892705718816354

McLaren (2020) Revealed: how McLaren is pioneering the use of sustainable composites in F1. https://www.mclaren.com/racing/team/natural-fibre-sustainable-composite-racing-seat/. Accessed 21 Sep 2020

Pickering KL, Efendy MGA, Le TM (2016) A review of recent developments in natural fibre composites and their mechanical performance. Compos Part A Appl Sci Manuf 83:98–112. https://doi.org/10.1016/j.compositesa.2015.08.038

Porsche_Motorsport (2020) 718 Cayman GT4 Clubsport. https://motorsports.porsche.com/international/en/article/718caymangt4clubsport. Accessed 3 Feb 2021

Stokke DD, Wu Q, Han G (2013) Introduction to wood and natural fiber composites: an overview. Wiley

Suddell BC (2008) Industrial fibres: recent and current developments. Proc Symp Nat Fibres 44:71–82

Suddell BC, Evans WJ (2005) Natural fiber composites in automotive applications. In: Mohanty AK, Misra M, Drzal LT (eds) Natural fibers, biopolymers, and biocomposites. Taylor & Francis, pp 231–259

Witayakran S, Smitthipong W, Wangpradid R, et al (2017) Natural fiber composites: review of recent automotive trends. Elsevier

Chapter 2
Natural Fiber Reinforcement Preparation

Abstract The textile reinforcement form (preform) plays an important role in determining the properties of the final composite/product. The preform formation process provides a precise control of the fiber architecture and orientation using a suitable textile manufacturing technique. While the techniques employed for preparing glass and carbon preforms are well-known, there is a gap in understanding how to prepare natural preforms for composite reinforcements. This chapter discusses the relevant preform preparation techniques and the resulting fiber architecture. Conventional preforms such as spun yarn, woven, knitted, nonwoven, braided, and comingled are illustrated and classified into one-, two- or three-dimensional reinforcements. Non-conventional preform formation techniques used in the automotive industry are also discussed, including, unidirectional tapes, pre-impregnated preforms, and power-Ribs. Finally, the structural parameters of each preform and their effect on the final composite properties are explained.

Keywords Natural fiber reinforcement · Preform · Woven · Nonwoven · Comingled · Unidirectional tape · Prepreg

The fibers architecture plays an important role in determining the properties of the final composite/product. The process of preforms preparation, regardless of the preparation methods, has many steps and factors that need optimization. This section will focus on reinforcement preparation techniques. Also, the section shows some non-conventional preforming techniques used in the automotive industry.

2.1 Conventional Preforms

Conventional preforms could be categorized from different points of views either from the preform structure point of view (1D, 2D, or 3D structures), or from the manufacturing technology point of view (nonwoven, knitting, weaving, or braiding) (Baley et al. 2019). This subsection discusses the conventional preforms categorized

© The Author(s), under exclusive license to Springer Nature Switzerland AG 2021
L. A. Elseify et al., *Manufacturing Automotive Components from Sustainable Natural Fiber Composites*, SpringerBriefs in Materials,
https://doi.org/10.1007/978-3-030-83025-0_2

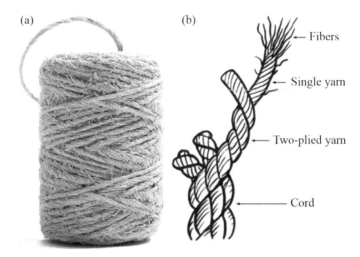

Fig. 2.1 a Natural fiber yarn and **b** schematic diagram showing the three types of yarn (Ahmadi et al. 2020; PalmFil.com 2020)

from the structure point of view then subcategorized based on the manufacturing technique.

2.1.1   1D Preforms

Fiber yarn is a continuous strand of fibers and could be single, plied, or cabled as shown in Fig. 2.1b. NF Yarn could be used in knitting or weaving to form reinforcing preforms. Figure 2.1a is a typical shape of natural fiber yarn or roving (Ahmadi et al. 2020; PalmFil.com 2020).

The most important parameter that affects yarn performance is the twist level. When twist level is excessively increased, the strength of the yarn is decreased due to the inclination of fibers. Additionally, high twisting level make the fibers tightly packed which hinders the resin penetration while processing (Goutianos et al. 2007; Ahmadi et al. 2020).

2.1.2   2D Preforms

Woven Woven preforms could have three types; plain, twill, or satin. The most commonly used weaving techniques with natural fiber reinforcements are the plain and twill weaves. Figure 2.2 shows some fiber preforms from the literature and the industry (Rayyaan et al. 2019; Bcomp 2020a; PalmFil.com 2020).

Fig. 2.2 Woven structures, **a** PalmFil plain-woven fabric, **b** flax hopsack fabric, and Bcomp **c** 2 × 2 and **d** 4 × 4 twill woven fabrics (Rayyaan et al. 2019; Bcomp 2020a; PalmFil.com 2020)

One of the preforms that are actually being used in the automotive industry is flax woven 2 × 2 twill fabric. Twill fabrics have good balance of moldability and structural integrity.

These fabrics were developed by Bcomp under the name of *ampliTex™ 5040*. Bcomp as a well-known manufacturer of flax reinforcements since 2003 creates other reinforcement structures. However, the *ampliTex™ 5040* is the only one used in the automotive industry.

Knitted Knitting is another type of 2D preforms which is formed by interlooping of yarns; however, this loop structure degrades the mechanical properties of the final composite. Knitting could be classified into warp knitting and weft knitting, yet, only the warp knitting is considered for making reinforcements due to the ability to lay-in straight thick rovings, while utilizing thin knitting yarns to stitch the straight rovings. Figure 2.3 shows a warp knitting unidirectional fabric made from twistless flax yarn (Rayyaan et al. 2019).

Nonwoven Nonwoven preforms depend on the formation of fiber web either by needle punching, thermal bonding, chemical bonding, or by stitching. Nonwoven preform formation process is considered economic and faster when compared to woven ones. Moreover, nonwoven mats have good mechanical properties and acceptable acoustic absorption properties (Baley et al. 2019).

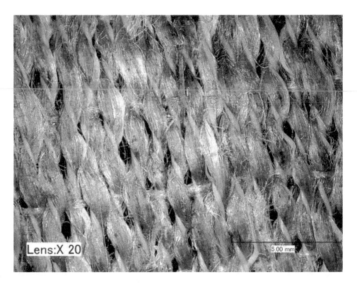

Fig. 2.3 Warp knitted unidirectional twistless warp-spun flax yarn (Rayyaan et al. 2019)

One of the widely used forms of natural fibers in the automotive industry is the nonwoven needle-punched laminates. Needle punch is a process that consolidates the fibers mechanically by the means of barbed needles. The barbed needles cause entanglements between the fibers by pushing the fibers in the web from one side to the other. The needle-punched mats are then transferred into composites using compression molding technique. A schematic diagram of the needle punch process is shown in Fig. 2.4a. Figure 2.4b shows PalmFil nonwoven fiber mat.

Another process developed by Johnson Controls Automotive called *Fibropur* is based on spraying a needle-punched mat made from flax, sisal, hemp, or kenaf with polyurethane (PU) and then heating the mat in an oven at 125 °C. Afterward, the composite is formed by applying pressure (Bledzki et al. 2002).

2.1.3 3D Preforms

3D woven preforms are made by introducing yarn in the z-direction. It provides higher through-thickness properties compared to 2D woven preforms. They can be classified as 3D orthogonal or 3D angular interlock. Figure 2.5 shows 3D angular warp interlock made from flax rovings (Baley et al. 2019; Lansiaux et al. 2019).

3D nonwoven preforms are formed by air-laying and hot-air-through bonding for achieving a thick high loft preform.

In 3D stitched preforms, a stack of 2D mats is stitched together. The advantage of this technique is the flexibility of changing the fibers orientation.

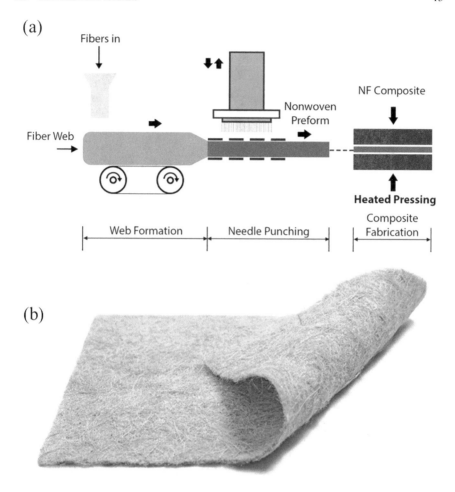

Fig. 2.4 **a** Needle punching process and **b** PalmFil nonwoven fabric (PalmFil.com 2020)

2.1.4 Comingled

Commingling technique is the process of combining both the natural fiber reinforcement and the thermoplastic matrix in fiber form, for making the dry preform. Commingling is classified into two categories: yarn and fabric. Commingled yarn could be blended, wrap-spun, or core-spun yarn as shown in Fig. 2.6. As can be seen in Fig. 2.6, hybrid fabric could be prepared from commingled yarn or from natural fiber yarn and thermoplastic yarn together (Ahmadi et al. 2020).

For composite manufacturing, pressure and heat must be applied to melt the thermoplastic matrix and impregnate the natural fibers.

Fig. 2.5 Manufacture of 3D-warp interlock from flax rovings (Lansiaux et al. 2019)

2.2 Non-conventional Preforms

2.2.1 Unidirectional Tapes

One of the non-conventional types of preforms is the unidirectional (UD) tape. UD form is achieved by aligning all the fibers in the same direction. The fibers are bound to each other mechanically (Fig. 2.7a) or by adhesive bond (Fig. 2.7b). Aligning the fibers has a very noticeable direct effect on the mechanical properties of the composite (Bar et al. 2019). One of the examples of the natural fibers unidirectional tapes is flax tapes. The flax tapes are made by water spraying the fibers then quickly drying them. This process causes the adhesion of fibers to each other. Flax tape is suitable

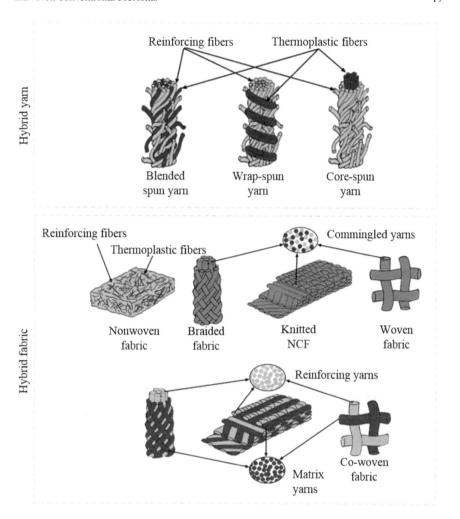

Fig. 2.6 Commingled natural fiber forms (Ahmadi et al. 2020)

for thin flat applications since they cannot be shaped into 3D shapes. The advantage of these tapes is that they are very light and provide good mechanical properties. A recent innovative product developed by Lineo is the Flaxpreg and Flaxpreg T-UD (Flax TapeTM), as shown in Fig. 2.7b, an impregnated flax fabric and flax tape, respectively (MaterialDistrict 2016; Mathijsen 2018). UD fibers render the strongest composites. However, it is not always easy to form stable UD fabrics. Moreover, handling UD fabrics is not flexible or versatile.

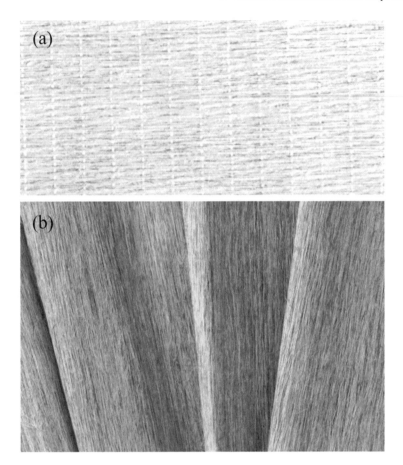

Fig. 2.7 **a** Bcomp UD ampliTex™ fabric and **b** Lineo flax tapes (MaterialDistrict 2016; Bcomp 2020a) ©Bcomp

2.2.2 Pre-impregnated Preforms

Pre-impregnated preforms, known as prepregs, could be defined as unidirectional or woven mats that are initially impregnated with resin. A prepreg is one of the reinforcement forms that are used to produce advanced composites. However, the use of prepregs in the automotive industry is not used extensively compared to the other preforms due to its high cost. Prepregs could be made with thermosets or thermoplastics. They are available as narrow or wide tapes with a width ranging between 15 cm and 1.5 m. The prepregs are made either by hot-melt impregnation or solvent impregnation processes. In the hot-melt impregnation, the polymer is heated to its melting point to facilitate the equal impregnation in the fiber mats. On the other hand, the solvent impregnation process is used with the polymers that need dissolving in a solvent prior to impregnation. For thermosets, the resin is used in a semisolid viscous

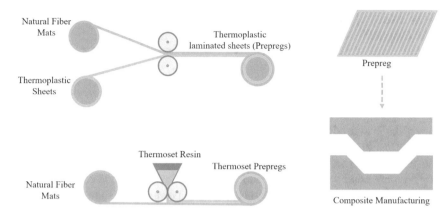

Fig. 2.8 Schematic diagram of thermoplastic and thermosets prepreg formation process

form at room temperature. Afterward, the composite can be manufactured by curing the thermoset resin under temperature and/or pressure. The most commonly used thermoset in prepregs is epoxy resin. However, for thermoplastics, the prepregs are made by a process called sheets lamination where thermoplastic sheets are used to laminate natural fiber mats to form prepregs. The produced prepregs are cooled down immediately after consolidation. Additionally, thermoplastic prepregs can be formed by water-slurry method. In the water-slurry method, the thermoplastic is supplied in the form of powder suspended in water (Campbell 2010; Litzler 2020). Figure 2.8 shows schematic diagrams of thermoplastic and thermoset prepregs formation processes. The composite is usually manufactured using compression molding process or manufactured into an autoclave or by vacuum bagging. For storage, the prepregs are usually covered by two protective sheets that are peeled off before composite manufacturing.

Natural fiber prepregs were manufactured by Composite Evolution using Evopreg technology and Bcomp's ampliTex™ and PowerRib™ flax fibers woven mats. The fibers are pre-impregnated with epoxy resin. The produced prepregs are lightweight with low density, thermal insulation, excellent noise, and vibration damping properties. Composite Evolution also produced carbon-flax hybrid Evopreg. The main aim of this hybrid prepreg is to replace most of the carbon fibers with flax fibers. Additionally, Composite Evolution also produces Evopreg PFC prepregs that are fire-retardant as well as Evopreg PFA which is pre-impregnated with bioresin (Evopreg 2021).

Another successful industrial attempt to manufacture natural fibers prepregs is BÜFA's unidirectional and woven thermoplastic BPREG. These prepregs are produced under the name EcoRein™. BÜFA uses hemp and flax fibers that are impregnated with PP. EcoRein™ are available with different thicknesses, weight, and dimensions. The unidirectional EcoRein™ are manufactured with 30:70 or 50:50 flax:PP. Additionally, they manufacture a totally biodegradable UD and woven prepreg that are made from 50% flax fibers and 50% PLA. The produced prepregs have modulus of elasticity between 9 and 13 GPa and flexural modulus between 6 and 9 GPa (BPREG 2021).

2.2.3   PowerRibs™

PowerRibs™ is an advanced invention by Bcomp. The idea of powerRibs™ is based on the formation of a lightweight grid that provides high stiffness at minimum weight. The powerRibs™ mats are designed to be used with thin shell mats like the ampliTex™ or any other woven natural fiber mats. They both act as the composite reinforcement. Figure 2.9 shows the structure of the powerRibs™. PowerRibs™ can be used with thermoplastics or thermosets. It can be used with thermoplastic with a base layer injection process to manufacture lightweight automotive interior parts. This process is highly efficient for large-scale production. Additionally, it can be used with thermosets with vacuum infusion, in autoclave or RTM to manufacture high-performance applications such as motorsport bodywork, aerospace, or space applications (Bcomp 2020b).

PowerRibs™ are made out of thick round flax yarn, with optimized fiber twisting, so that it is not high that it affects the mechanical strength or not very low that the fibers lose their longitudinal stiffness. In general, twisting facilitates the handling of fibers. However, twisting will cause uneven loading of fibers against tension, as mentioned earlier. Moreover, twisting will lead to stiffness reduction as a result of sideways loading on fibers. PowerRibs™ are zero and 90° nets sewn together using thin polyester yarn. This binding technique gives the powerRibs™ flexibility to be shaped into complex molds as shown from Fig. 2.10. Also what adds to powerRibs™ is that the composites reinforced with them could be manufactured by applying pressure in an autoclave either using vacuum bag or compression molding (Mathijsen 2018). Figure 2.11 shows how powerRibs™ are layered with ampliTex™ to manufacture composite panels in flexible shapes (Bcomp 2020a, b).

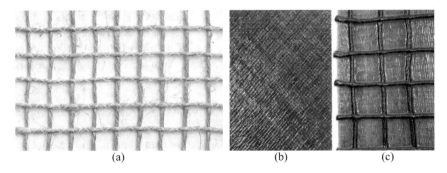

(a)   (b)   (c)

Fig. 2.9 a Bcomp flax powerRibs™ mat, and Bcomp powerRibs™ and ampliTex™ composite showing **b** back and **c** front sides (Bcomp 2020b) ©Bcomp

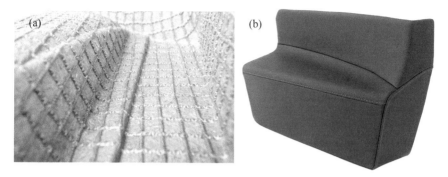

Fig. 2.10 a Bcomp powerRibs™ thermoplastic **b** automotive interior panel covered with leather ©Bcomp

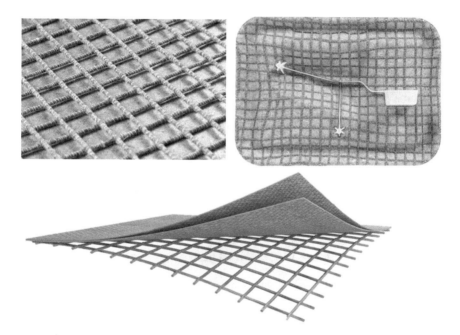

Fig. 2.11 Close up look of Bcomp powerRibs™ and ampliTex™ (Bcomp 2020a, b) ©Bcomp

In conclusion, natural fibers could be made into any of the previously mentioned structures. However, choosing which fiber architecture to use depends on the final composite properties required and if the composite is intended to be used in structural or non-structural applications. Each preform structure has its advantages and drawbacks. However, for the automotive industry, nonwoven preforms and unidirectional tapes are favored. The nonwoven preforms are favored because it is a fast cost-effective process. Also, it has good properties in all directions due to the randomness of the fiber distribution. Additionally, using commingling techniques will make the manufacturing process much faster especially in the making of door panels. On the other hand, unidirectional tapes and woven preforms similar to Bcomp's might be favored in structural parts and when high volume fraction is needed.

References

Ahmadi MS, Dehghan-Banadaki Z, Nalchian M (2020) Date palm fiber preform formation for composites. In: Midani M, Saba N, Alotman O (eds) Date palm fiber composites, 1st edn. Springer Singapore, pp 93–117

Baley C, Gomina M, Breard J et al (2019) Specific features of flax fibres used to manufacture composite materials. Int J Mater Form 12:1023–1052. https://doi.org/10.1007/s12289-018-1455-y

Bar M, Alagirusamy R, Das A (2019) Development of flax-PP based twist-less thermally bonded roving for thermoplastic composite reinforcement. J Text Inst 110:1369–1379. https://doi.org/10.1080/00405000.2019.1610997

Bcomp (2020a) ampliTex™. http://www.bcomp.ch/en/products/amplitex

Bcomp (2020b) PowerRibs™. http://www.bcomp.ch/en/products/powerribs. Accessed 18 May 2020

Bledzki AK, Sperber VE, Faruk O (2002) Natural and wood fibre reinforcement in polymers. Macromolecules 13:3–144

BPREG (2021) BPREG Flax UD Tapes. http://bpreg.com/. Accessed 15 Mar 2021

Campbell FC (2010) Structural composite materials. ASM International

Evopreg (2021) Composite evolution

Goutianos S, Peijs T, Nystrom B, Skrifvars M (2007) Textile reinforcements based on aligned flax fibres for structural composites. Compos Innov 1–13

Lansiaux H, Corbin A-C, Soulat D et al (2019) Identification du comportement mécanique de Tissu 3D interlock chaine À base de Mèches de Lin. Rev Compos Des Mater Av 29:73–81. https://doi.org/10.18280/rcma.290111

Litzler (2020) Thermoplastic Prepreg Systems. The two,web is saturated with the slurry... More. https://www.calitzler.com/prepreg-systems/thermoplastic-prepreg-systems/#:~:text=The rmoplasticPrepregSystems. Accessed 23 Jun 2021

MaterialDistrict (2016) FlaxTape. https://materialdistrict.com/material/flaxtape/. Accessed 23 Jan 2020

Mathijsen D (2018) The renaissance of flax fibers. Reinf Plast 62:138–147. https://doi.org/10.1016/j.repl.2017.11.020

PalmFil.com (2020) PalmFil. http://www.palmfil.com/. Accessed 29 Aug 2020

Rayyaan R, Kennon WR, Potluri P, Akonda M (2019) Fibre architecture modification to improve the tensile properties of flax-reinforced composites. J Compos Mater. https://doi.org/10.1177/0021998319863156

Chapter 3
Natural Fiber Composite Fabrication for the Automotive Industry

Abstract Selecting the proper composite fabrication technique in the automotive industry is key to fulfilling the performance criteria and economic feasibility of the manufactured parts. While there are several standard techniques that are commonly used with carbon and glass fibers, only few of those techniques are suitable for use with natural fiber reinforcements. This chapter will discuss the different natural fiber composite (NFC) manufacturing techniques used in the literature as well as those techniques used by leading automotive parts manufacturers. Compression molding, resin transfer molding, and vacuum-assisted resin transfer molding techniques are described with details on processing parameters and procedures. Hand layup technique as well as extrusion and injection molding is also discussed. Comparison between processing parameters of NFC with thermoplastic and thermoset matrices is summarized in addition to comparison of the mechanical properties of NFC manufactured using selected techniques.

Keywords Natural fiber composite · Compression molding · RTM · VRTM · Hand layup · Injection molding

Composites are generally manufactured in the automotive industry using compression molding, resin transfer molding (RTM), filament winding, injection molding, pultrusion, and layup process. Likewise, natural fiber-reinforced composites in the automotive industry are manufactured using the same techniques except that filament winding, and pultrusion are not widely used with natural fibers. Table 3.1 shows a comparison between the common manufacturing processes used in the automotive industry. This section will discuss the different NFC manufacturing techniques used in the literature and by the leading automotive parts manufacturers.

© The Author(s), under exclusive license to Springer Nature Switzerland AG 2021
L. A. Elseify et al., *Manufacturing Automotive Components from Sustainable Natural Fiber Composites*, SpringerBriefs in Materials,
https://doi.org/10.1007/978-3-030-83025-0_3

Table 3.1 Comparison between manufacturing processes used in the automotive industry

Process	Advantages	Disadvantage
Compression molding	• Used with short or long fibers • Good surface finish • Low waste of material • Low-cost process • Short setup time • Fiber content can be easily controlled	• Slow production process • Not suitable for thin parts • Large parts need a huge presser
Resin transfer molding (RTM)	• High fiber volume fractions can be obtained with low void content • Excellent surface finish • Less labor cost • Very suitable for complex geometries	• Expensive process • Applied pressure cause fibers mobility leading to fiber wash • Limited to manufacturing of small parts
Vacuum-assisted resin transfer molding (VARTM)	• Lower tooling cost than RTM • Manufacturing of large components • Flexible	• Difficult to use with high viscosity resins • Dry spots formation • More complex than RTM • Surface finish is not as good as RTM process
Hand layup	• Simplicity • Low tooling cost • Used in manufacturing of complex parts	• Process accuracy fully depend on the skills of worker • Low viscosity resins must be used to ensure resin impregnation • Some voids can be produced • Low fiber content
Extrusion	• Low production cost • Short setup time	• Only used for manufacturing of uniform cross-sectional parts • Limited to short fibers
Injection molding	• Low cost • Suitable for complex parts • Very precise process	• Mostly used with short fibers

3.1 Compression Molding

Compression molding is a process that depends on heat and pressure to reshape the material. In compression molding, working with thermoplastics is much easier than working with thermosets. Compression molding could be classified into sheet molding compound (SMC) and bulk molding compound (BMC). The difference between the two processes is that in BMC the charge is in the form of fibers–matrix mixture; but, in SMC the charge is in the form of sheets where two layers of polymers are pressed together with the reinforcement in between. Figure 3.1 is a schematic diagram of SMC and BMC. Sheet molding compound (SMC) is a frequently used technique in the automotive industry. It is mainly for thermosets where it is used in the manufacturing of bumpers, trunk covers, and spoilers (Bledzki et al. 2002).

The compression molding could be used with short or long fibers and in woven or nonwoven forms. In this process, the composite constituents are prepared first; then, the compression molding parameters; temperature, duration, and pressure are determined and set. Figure 3.2 shows the steps of the compression molding process.

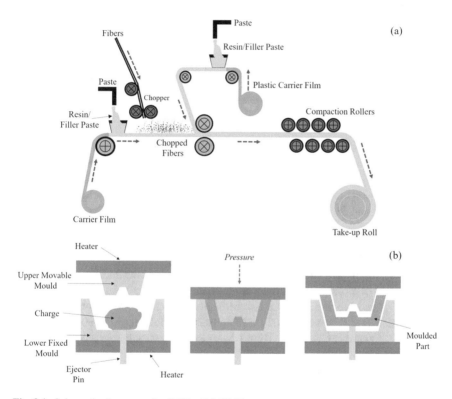

Fig. 3.1   Schematic diagrams of **a** SMC and **b** BMC

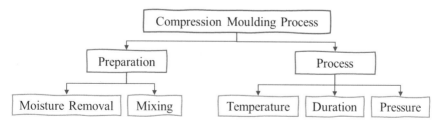

Fig. 3.2   Compression molding process steps

Compression molding process is one of the extensively used processes with natural fibers in the literature and mostly with thermosets. This could be due to the fact that thermosets are much easier to process than thermoplastics. The preparation of composite constituents could be in the form of preform preparation, matrix–fiber mixing, and/or fiber drying to remove any excess moisture. After analyzing several studies, it was noticed that the most commonly used drying temperatures were 70, 80, and 100 °C (Sbiai et al. 2008; Mulinari et al. 2011; Abdal-hay et al. 2012; Jang et al. 2012; Mir et al. 2013; Mohanty et al. 2014; Rohen et al. 2015; Duquesne et al. 2015; Mahdi et al. 2015; Neher et al. 2016; Ali and Alabdulkarem 2017; Papa et al. 2017; Mohanty 2017; Li et al. 2019; Mazzanti et al. 2019; Tataru et al. 2020; Kharrat et al. 2020). Moreover, according to a recent review study (Jaafar et al. 2019) the optimum drying temperature to remove excess moisture was found to be 80 °C. But in order to claim that a certain parameter level is the optimum, the effect of several levels must be studied. However, after reviewing the mentioned studies it was noticed that these studies did not attempt to use other temperature levels to claim that 80 °C is the optimum temperature.

As for determining the most frequently used drying combination (duration and temperature), it was found that the drying durations were not even within the same range for the same drying temperature. Therefore, it is hard to conclude a single drying combination that worked perfectly with natural fibers. Some researchers used very long durations exceeding 2 days (Abu-Sharkh and Hamid 2004; Mirmehdi et al. 2014; Alajmi and Shalwan 2015; Ali and Alabdulkarem 2017; Masri et al. 2018; Prabhu et al. 2019; Abdullah et al. 2020; Syaqira et al. 2020). While others limited the drying duration to 1 or 2 h (Sbiai et al. 2008; Mir et al. 2013; Suresh Kumar et al. 2014; Li et al. 2015b, a; Mahdi et al. 2015; de Araujo Alves Lima et al. 2019; Xu et al. 2019), the most frequently used drying duration was the 24 h (Al-Kaabi et al. 2005; Haque et al. 2010b; Mulinari et al. 2011; Abdal-hay et al. 2012; Dehghani et al. 2013; Shalwan and Yousif 2014; Mohanty et al. 2014; Duquesne et al. 2015; Neher et al. 2016; Zadeh et al. 2017; Mohanty 2017; Motaung et al. 2017; Mazzanti et al. 2019; Azammi et al. 2020). However, considering sustainable production, 24 h is considered a very long duration.

Moving to the process molding parameters (temperature, pressure, and duration). The molding temperature should not exceed the degradation temperature of fibers (i.e., below 250 °C) and should be high enough to ensure adequate fibers wetting. A study was made by Rokbi et al. (2020) to study the effect of the processing parameters on the mechanical properties of plain-woven jute fabric reinforced with PP. Rokbi et al. made eight samples four of which have constant pressure of 2 MPa and varying temperature levels (190, 210, 230, 250 °C). While the other four samples have constant temperature of 200 °C and varying pressure levels (1, 2, 3, 4 MPa), it was noticed that as the molding temperature was increased, the tensile strength of the composite was decreased. This could be due to the degradation of jute fibers when exposed to high-temperature levels as 250 °C. On the other hand, when the pressure was increased the tensile strength values were significantly increased indicating that the increasing pressure plays an important role in improving the fiber–matrix interfacial bond. Additionally, high consolidation pressure results in higher fiber volume fraction. As for the elongation at break %, the value was reduced when either the temperature or the pressure was increased. Finally, the modulus of elasticity results showed that when the temperature was increased the modulus of elasticity increased till it reached certain value then it decreased again. However, when the applied pressure was increased the modulus of elasticity was increased. The authors concluded that the effect of the applied pressure is more significant than the effect of the temperature. Several studies were reviewed from the literature to determine the most commonly used processing parameters with natural fibers. Tables 3.2 and 3.3 show the compression molding parameters obtained from the literature for natural fiber composites with thermoplastics and thermosets, respectively.

Finally, the last affecting factor is the holding duration. Using short durations caused some defects in the produced composites due to insufficient time to apply pressure and temperature. It is very important to plan the process duration to make sure that the resin has enough time to completely wet the fibers especially if fibers are tightly packed. However, it can be concluded that there is no correct exact value for each parameter but a correct combination of the three parameters. Choosing the right molding parameter is essential for producing high-strength reliable composites (Jaafar et al. 2019; Rokbi et al. 2020).

Consequently, after reviewing the work done in the literature for thermoplastic composites and the data presented in Table 3.2, it was noticed that high-pressure and temperature values were used with shorter durations and vice versa, supporting the previously discussed point. The temperature levels used with thermoplastics were between 130 and 250 °C. However, with thermoplastic starch a temperature between 130 and 160 °C was used (Ochi 2006; Ibrahim et al. 2014, 2017; Saleh et al. 2017). On the other hand, PP was used with temperature above 170 °C (Lai et al. 2005; Shibata et al. 2006; Asumani et al. 2012; Mir et al. 2013; Bar et al. 2018; Ovlaque et al. 2020). As for the molding pressure used, it was noticed that the value of the pressure used with thermoplastics was much higher than that used with thermosets. This is to force the thermoplastic melt to penetrate between the tightly packed fibers and to improve the consolidation between fibers and matrix and to remove the excess resin. Finally, the molding duration for thermoplastics was mostly in the range of few minutes. As for the processing parameters of thermoset composites, it was noticed that most of the composites, as presented in Table 3.3, were manufactured at room temperature with very low-pressure values compared to thermoplastic samples. Generally, most of the thermosets are cured at room temperature, and to increase the curing rate, the temperature is increased. A study was made in 2015 to test the effect of the processing parameters on the void content, showed that increasing curing pressure had a positive effect on reducing the void content, hence improving the tensile strength of the samples (Li et al. 2015a). However, increasing the molding temperature was not effective as increasing pressure. As shown in Table 3.3, most of the works done have used longer durations (more than 24 h) while working at room temperature. However, some authors used comparably short durations (1–2 h) but with temperatures higher than 50 °C (Jacob et al. 2006; Jiang et al. 2009; Muralidhar et al. 2012; Muralidhar 2013; Li et al. 2015a; Rohen et al. 2015; Pushparaja et al. 2017; Boutin et al. 2020; Mishra et al. 2020). After analyzing the compression molding

Table 3.2 Compression molding process parameters for NF and thermoplastics composites from the literature

Fiber	Matrix	Fibers form	Fiber (wt%)	Temperature (°C)	Pressure (MPa)	Duration (min)	References
DP spadix stems	Corn starch	Chopped	0, 20, 40, 50, 60, 80	160 ± 3	5	30	Ibrahim et al. (2014)
Flax	Corn starch	Chopped	50	160 ± 3	5	30	Ibrahim et al. (2014)
DP spadix stems	Corn starch	Chopped	50	160	5	30	Ibrahim et al. (2017)
Flax	Corn starch	Chopped	50	160	5	30	Ibrahim et al. (2017)
Banana	Corn starch	Chopped	50	160	5	30	Ibrahim et al. (2017)
Bagasse	Corn starch	Chopped	50	160	5	30	Ibrahim et al. (2017)
DP leaflets	Recycled EPS	Milled	70, 75, 80	–	1	10	Masri et al. (2018)
DP mesh	Corn Starch	Short	0, 20, 50, 70	130	5 tonnes	60	Saleh et al. (2017)
Sisal	PP	Nonwoven	50, 60, 70, 80	–	13.8	1.5	Prajwal et al. (2019)
Bamboo	PP	Lamellae	–	200	5 min 0 tonne, 1 min 4 tonnes, 1 min 5 tonnes	5 min 0 tonne, 1 min 4 tonnes, 1 min 5 tonnes	Ovlaque et al. (2020)
Bamboo	PLA	UD	18.37	180	6	–	Roy Choudhury and Debnath (2020)
Abaca	Starch-based resin	UD	30–70	130	10	10	Ochi (2006)
Coir	PP	Short	0, 10, 15, 20	170	30 KN	20	Mir et al. (2013)
Coir	PP	Chopped	0, 5, 10, 15, 20, 25	135–170	7–20	–	Lai et al. (2005)

(continued)

Table 3.2 (continued)

Fiber	Matrix	Fibers form	Fiber (wt%)	Temperature (°C)	Pressure (MPa)	Duration (min)	References
Coir + Sisal	PP	Commingling	13.06, 20.89, 31.2	210	0.5	9	Arya et al. (2015)
Kenaf	HDPE	Chopped	0, 3.4, 8.5, 8.5, 17.5	155	98	–	Salleh et al. (2014)
Kenaf	PP	Nonwoven	20, 25, 30, 35	250	7.5	15	Asumani et al. (2012)
Kenaf	PP	Chopped	20–65	230	10	–	Shibata et al. (2006)
Kenaf	PLA	Chopped	0, 10, 30, 50, 70	200	0.7	5	Lee et al. (2009)
Jute	PLA	Plain woven, UD	–	180	50 KN	10	Khan et al. (2016)
Ramie	HDPE	Powder	30, 40, 50	130	0.62	60	Banowati et al. (2016)
Flax	PP	Commingling	40, 50, 60	190	0.8	5	Bar et al. (2018)
Flax	Potato starch	4 × 4 hopsack	50–60	180	1	5	Duquesne et al. (2015)
Flax	PBAT	4 × 4 plain, 2 × 2 twill	–	150	10.34	5	Phongam et al. (2015)

Table 3.3 Compression molding process parameters for NF and thermosets composites from the literature

Fiber	Matrix	Fiber wt%	Temperature (°C)	Pressure (MPa)	Duration (h)	Post Curing	References
DP leaflets	Epoxy	18	Room temp	0.2	24	2 h 60 °C, 1.5 h 90 °C, 1 h 120 °C, 30 min 150 °C	Sbiai et al. (2008)
DP leaflets	Epoxy	31.4	Room temp	0.1	24	6 h at 80 °C	Abu-Sharkh and Hamid (2004)
DP mesh	Epoxy + Graphite	35	Room temp	–	24	24 h at 50 °C	Alajmi and Shalwan (2015)
DP mesh	Epoxy + Graphite	35	Room temp	–	24	24 h at 50 °C	Shalwan and Yousif (2014)
DP mesh	Unsaturated Polyester	6, 7, 8, 9, 10	Room temp	–	24	2 h at 100 °C	Al-Kaabi et al. (2005)
Sisal + Glass	Epoxy	5, 10, 15, 20, 25	Room temp	147		–	Prabhu et al. (2019)
Sisal	Epoxy	40, 50, 60	Room temp	–	24	–	Anidha et al. (2020)
Sisal	Epoxy	0, 5, 10, 15	Room temp	–	24	2 h at 120 °C	Srisuwan and Chumsamrong (2012)
Sisal	Natural rubber sheets	–	150	–	8 min	–	Jacob et al. (2006)
Sisal	Unsaturated Polyester	–	Room temp	17	24	–	Senthilkumar et al. (2017)
Sisal + Cotton	Unsaturated Polyester	10, 20, 30, 40, 50	Room temp	0.15	4	–	Sathishkumar et al. (2017b)
Sisal	Unsaturated Polyester, Epoxy	0, 10, 20, 30	Room temp, then 80	7	24 then 4	2 h at 110 °C then 15 min at 140 °C	Jiang et al. (2009)
Sisal + Kenaf	Bio-epoxy	20	Room temp	–	24	24 h at 80 °C	Yorseng et al. (2020)
Sisal + Jute	Epoxy	30	80	Hydraulic press	6	–	de Araujo Alves Lima et al. (2019)

(continued)

Table 3.3 (continued)

Fiber	Matrix	Fiber wt%	Temperature (°C)	Pressure (MPa)	Duration (h)	Post Curing	References
Sisal + Coir	Epoxy	25	80	–	4	–	Pushparaja et al. (2017)
Coir	Epoxy	3, 9, 15	Room temp	0.5–4 kg	24–48	–	Romli et al. (2012)
Coir	Epoxy	47–48	Room temp	3.92	24	–	Suresh Kumar et al. (2014)
Kenaf	Epoxy	50	Room temp	27.5	5 min	–	Hanan et al. (2018)
Kenaf + Banana	Unsaturated Polyester	40	Room temp	45 kg	24	1 h at 50 °C	Alavudeen et al. (2015)
Kenaf + Bamboo + Jute	Unsaturated Polyester		Room temp	–	24	2 h at 100 °C	Hojo et al. (2014)
Kenaf	Kevlar and Epoxy	51	Room temp	–	24	–	Yahaya et al. (2014)
Kenaf	Epoxy	28	Room temp	–	48	–	Alkbir et al. (2016)
Hemp	Vinyl ester	44–45	Room temp	4.36 kPa	24	4 h at 80 °C	Misnon et al. (2016)
Flax	Epoxy	60	90, 100	0.5, 1, 1.5	20, 30, 40 min	–	Li et al. (2015a)
Flax	Epoxy	–	50	0.3	2	1 h at 120 °C	Muralidhar et al. (2012)
Flax	Epoxy	21.17–38.44	50	0.3	2	1 h at 120 °C	Muralidhar (2013)
Flax	Epoxy	30.78, 38.44	Room temp	–	24	8 h at 60 °C	Di Bella et al. (2010)
Coir + Cotton + Sugar Cane	Green Epoxy	10, 15, 20	80, curing at room temp	10	0.5	–	Hassan et al. (2020)
Jut + Flax + Glass	Epoxy	34, 35, 36, 37	100	0.284	1.5	–	Mishra et al. (2020)
Flax	Epoxy system	40	50	0.5	1	1 h at 150 °C	Boutin et al. (2020)
Jute + Glass	Dicarboxylic acid	–	Room temp	–	2	–	Ouarhim et al. (2020)

parameters for thermoplastic and thermoset composites, and the data presented in Tables 3.2 and 3.3, it was noticed that usually the molding duration for thermoplastic composites is much less than that for thermosets. Thermoplastics processing durations are in the matter of minutes, whereas thermosets processing durations are in the range of hours. Comparing the molding pressure used with thermoplastics and thermosets, it was observed that thermoplastics need comparably high-pressure values than thermosets. Moreover, the same was observed with the temperature levels used. Thermosets already exist in viscous liquid forms while thermoplastics need high temperature to reach their melting point.

The mechanical properties of composites were reviewed and to have a fair comparison between the works done in the literature, the composites reinforced with the same fiber weight fraction were compared to each other. It was found that there are many factors that affected the final mechanical properties such as fiber volume fraction, processing parameters, and type of fiber. Generally, increasing the fiber weight fraction should contribute to the mechanical strength of the composite. However, in some cases, when the interfacial bond between fibers and matrix is weak, the strength of the composite dropped when fiber wt% increased. Usually, in these cases the neat composite would exhibit higher tensile strength than reinforced composites. The processing parameters could also affect the composite properties since the composite strength is dependent on the pressure, duration, and temperature levels used. As mentioned before, increasing the processing pressure reduced the voids inside the composite, hence increasing strength (Li et al. 2015a). Moreover, choosing the appropriate temperature level is very important. It should be optimized such that it is not very high that the fibers degrade or very low that the polymer is not totally melted. Finally, the type of natural fiber used could affect the final properties of the composites since their properties are not the same (Elseify and Midani 2020).

The effect of fiber weight fraction of fibers on the mechanical properties of thermoplastic composites was reviewed. It is expected that as the fiber wt% increases the strength of composite will increase since fibers are the load-bearing element in the composite (Banowati et al. 2016; Bar et al. 2019; Prajwal et al. 2019). On the other hand, other researchers found that increasing the fiber wt% beyond a certain value had a negative effect on the mechanical properties (Lee et al. 2009; Ibrahim et al. 2014, 2017). This is probably due to weak interfacial bond between fibers and matrix. This problem by applying to appropriate pressure for sufficient duration to ensure the wetting of all fibers, hence strengthening the bond between fibers and matrix. The tensile strength of HDPE reinforced ramie composite was found to be 31.31 MPa (Banowati et al. 2016). Kenaf reinforced PP composite was made with 30% fiber wt% and had tensile strength of 55 MPa and modulus of elasticity of 1.8 GPa (Asumani et al. 2012). Ibrahim et al. (2017) manufactured corn starch composites reinforced with flax, banana, bagasse, and date palm (DP) spadix stem fibers, and the mechanical properties of the manufactured composites were 41 ± 7.1 MPa, 25.4 ± 3.8 MPa, 29.8 ± 3.1 MPa, and 28.2 ± 2.9 MPa, respectively. Coir fiber-reinforced PP composite was reinforced with 0, 10, 15, and 20% fiber wt%. The results showed that the reinforced composites had mechanical properties worse than the neat composite in most of the cases with tensile strength dropping from 32 to 26 MPa when reinforced with 10% and 20% fiber wt%, respectively (Mir et al. 2013). It is assumed that in this case the fibers are regarded as inclusion rather than a reinforcement.

As for thermoset composites, different studies showed that the composite strength decreased when the fiber wt% decreased (Sathishkumar et al. 2017b; Prabhu et al. 2019; Yorseng et al. 2020). Unsaturated polyester composite reinforced with 40 wt% sisal fibers at 0.15 MPa for 4 h the flexural strength of sisal composite was 275 MPa, with flexural modulus of 4 GPa, tensile strength of 65 MPa, and Young's modulus of 0.55 GPa (Sathishkumar et al. 2017b). Coir-reinforced composites had tensile strength of nearly 58.6 MPa and flexural strength of 76.8 MPa when reinforced with 50% fiber wt% (Suresh Kumar et al. 2014). Furthermore, 50 wt% kenaf reinforced epoxy composite had tensile strength of 84.32 MPa with flexural strength and flexural modulus of 95 MPa and 7.5 GPa, respectively (Hanan et al. 2018). As for flax fiber composites, 60% reinforced epoxy composites exhibited tensile strength and modulus of elasticity equal 205.56 ± 8.14 MPa and 22.67 ± 0.47 GP, respectively (Li et al. 2015a). However, lower fiber volume fraction rendered lower tensile strength results in the range between 80 and 150 MPa (Di Bella et al. 2010; Muralidhar et al. 2012; Muralidhar 2013; Boutin et al. 2020; Hassan et al. 2020).

Compression molding could be regarded as one of the most suitable manufacturing processes to be used with natural fibers in making automotive components. The procedures are not complicated and could be easily controlled.

3.2 Resin Transfer Molding (RTM) and Vacuum-Assisted Resin Transfer Molding (VARTM)

RTM is a technique used with thermosets to form fiber-reinforced composites. This technique is based on the transfer of thermoset resin from an external source to dry fiber mats or preforms (Hassanin et al. 2020). The reinforcements could either be unidirectional or random depending on whether the final product is a structural or non-structural part. This process has been used with flax, sisal and hemp, and other natural fibers (Bledzki et al. 2002; Summerscales and Grove 2013). As shown in Fig. 3.3 the dry reinforcements are placed in a mold; afterward, the resin is allowed to enter the mold from one side and leave from the other side after wetting all fibers/fabric.

Vacuum-assisted resin transfer mold (VARTM) is a modified RTM process where the upper mold half of TM is replaced by a vacuum bag. This technique usually creates low void composites. In this technique, the reinforcements are laid on the mold then enclosed by a bagging material. This bagging material has two ports: a resin inlet port connected to the resin supply container and a resin outlet port connected to vacuum pump as shown in Fig. 3.4. After the resin has completely infused through the preform, the supply of resin is closed, and the composite is left to cure under

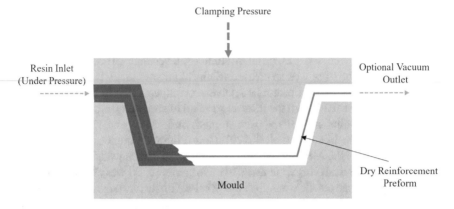

Fig. 3.3 Schematic diagram of RTM process

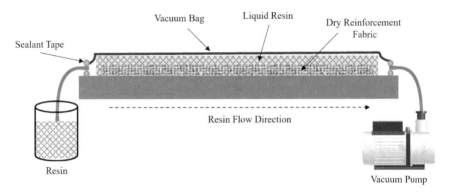

Fig. 3.4 Schematic diagram of VARTM process

vacuum pressure (Hassanin et al. 2020). This process is sometimes referred to as vacuum-assisted resin infusion, vacuum bagging, or resin infusion.

RTM and VARTM processes were used with natural fibers in the literature, and successful composite fabrication attempts were made by many researchers. Table 3.4 summarizes and shows the mechanical properties of natural fiber composites manufactured using RTM and VARTM. In VARTM, the vacuum pressure used in most of the cases was between 0.2 and 0.4 MPa (Santulli 2000; Wazzan 2006; Li et al. 2015b). However, some researchers did not mention the exact processing parameters used. Flax epoxy composites were cured at 80 or 150 °C for 2 h, or at room temperature for 24 h (Di Bella et al. 2010; Rayyaan et al. 2019; Tataru et al. 2020). However, when flax was used with soybean oil, it was cured at room temperature for 3–5 h (O'Donnell et al. 2004). The impact absorption energy of jute and flax composites manufactured using RTM was found to be 27 ± 3 kJ/m^2 and 32.6 kJ/m^2, respectively (Santulli 2000; Bertomeu et al. 2012).

Table 3.4 Mechanical properties of natural fiber composites manufactured using RTM and VARTM

Process	Fiber	Fiber form	Fiber (%)	Matrix	σ (MPa)	E (GPA)	References
RTM	DP mesh	Knitted	60 vol%	Unsaturated polyester	75	4.22	Wazzan (2006)
	DP mesh	Unidirectional	60 vol%	Unsaturated polyester	85	4.17	Wazzan (2006)
	Flax	Knitted	30 vol%	Unsaturated polyester	143	14	Goutianos et al. (2007)
	Sisal	Unidirectional	50 wt%	Epoxy	157.89	–	Li et al. (2015b)
	Hemp	Braided	–	Epoxy	62.5	4.581	del Borrello et al. (2020)
VARTM	Jute	Plain	60 wt%	Unsaturated polyester	60 ± 3.5	5.6 ± 0.2	Santulli (2000)
	Flax	Bidirectional fabric	–	Epoxy	80	1.75	Di Bella et al. (2010)
	Flax	Fabric	–	Epoxy	80	1.2	Bertomeu et al. (2012)
	Flax	Hopsack	40 vol%	Epoxy	73.32 ± 7.29	7.91 ± 0.26	Rayyaan et al. (2019)
	Flax	Unidirectional	40 vol%	Epoxy	103.85 ± 2.25	8.87 ± 0.33	Rayyaan et al. (2019)
	Flax	Nonwoven	40 vol%	Epoxy	110.15 ± 5.06	12.5 ± 1.18	Rayyaan et al. (2019)
	Flax	Woven	33%	Epoxy	87.64	–	Oliver-Borrachero et al. (2019)
	Jute	Woven	27.3%	Epoxy	50.32	–	Oliver-Borrachero et al. (2019)
	Flax	Twill 2 × 2	–	Epoxy	140	–	Tataru et al. (2020)

σ Tensile strength, E Young's modulus

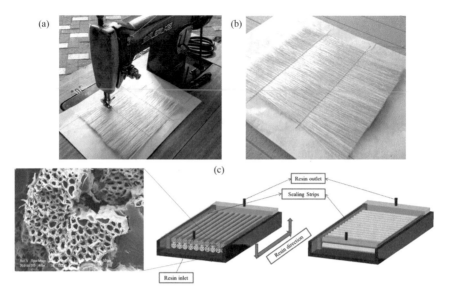

Fig. 3.5 **a** Sewing fabrication method, **b** unidirectional sisal fiber fabrics, and **c** composites with and without resin penetration (Li et al. 2015b)

RTM and VARTM are very suitable for the automotive industry and could be used successfully with natural fibers. These two processes could be used for high-performance applications. However, they are expensive as compared to the other manufacturing techniques.

An interesting work was made by Li et al. in (2015b). They investigated the effect of epoxy resin penetration into the lumens of sisal fibers. The fibers were sewn together using a sewing machine. The sisal was sewn first on a paper then this paper was torn off as shown in Fig. 3.5a, b. Two composite samples were made: one prepared while allowing the resin to penetrate inside the lumens and the other without using sealing tapes as shown in Fig. 3.5c. It was concluded that resin penetration improved the tensile strength and flexural strength. Moreover, it was noticed that the penetrated resin delayed the composite crack propagation. Additionally, the penetrated resin improved the impact strength and the water resistance property of the composites since almost all the fiber lumens were filled with resin; hence no place for water.

3.3 Hand Layup

Hand layup is a widely used composite manufacturing technique where the resin or the matrix impregnates the fiber by hand and further consolidation is achieved using a hand roller. First, dry reinforcement fabric is laid in the mold then liquid resin is poured over the fibers followed by hand consolidation using steel roller. The sample is then left to cool at room temperature (Hassanin et al. 2020). Figures 3.6 and 3.7 show a schematic diagram of hand layup technique and jute/epoxy composite manufactured using hand layup technique, respectively.

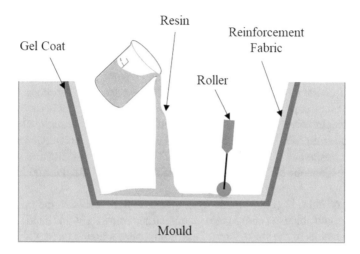

Fig. 3.6 Schematic diagram of hand layup

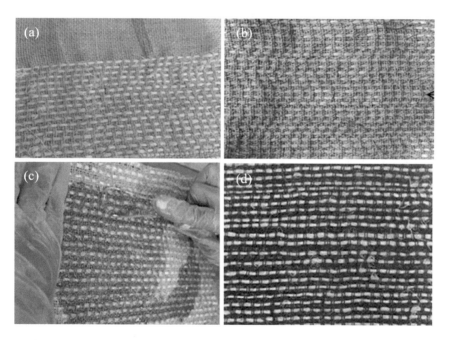

Fig. 3.7 Jute/sisal hybrid fabric **a** in production and **b** final product. Manufacturing of composite using **c** hand layup technique and **d** final product (de Araujo Alves Lima et al. 2019)

This process was found to be used mostly with epoxy resin. The fibers were used in the woven and the nonwoven forms. Some of the work published on composites manufactured using layup was reviewed. Jute/sisal epoxy composite with 30 vol% had shear strength of 26.56 ± 5.1 MPa, and flexural strength of 131.58 ± 22.9 MPa (de Araujo Alves Lima et al. 2019). Sisal/ kenaf hybrid epoxy composite reinforced with 65% fibers had tensile strength of 23.74 MPa, flexural strength of 1 MPa, and impact energy of 6.1 J (Palani Kumar et al. 2017). Another attempt was made by (Yousif et al. 2012) to manufacture kenaf epoxy composite. The composite was reinforced with 38 – 41 vol% fibers. The final composite had flexural strength of 300–350 MPa and 7 GPa flexural modulus. Polyester as a matrix was used by Mekonnen and Mamo (2020). The matrix was reinforced with jute, bamboo, and a mix of both. In all cases, the fiber weight percentage was kept at 30 wt%. The tensile strength values of jute and bamboo composites were 46.61 MPa, and 53.87 MPa, respectively. However, the hybrid composite had tensile strength of 72.03 MPa. Finally, the flexural strength of jute, bamboo, and the hybrid composites were 114 MPa, 133.9 MPa, and 131.5 MPa, respectively.

Goutianos et al. in (2007) made a comparative study between the properties of flax/unsaturated polyester composites manufactured using RTM and hand layup technique. The results of the study showed that composites manufactured with RTM had higher mechanical properties than those manufactured with hand layup technique (Goutianos et al. 2007).

Layup technique works well with natural fibers. However, this technique is not practical in the automotive industry because of the repeatability and quality consistency, in addition to the extensive manual work. Additionally, concerning large-scale production, these two processes may not be very suitable when compared to the other manufacturing techniques discussed in this review.

3.4   Extrusion and Injection Molding

In the automotive industry, extrusion is used especially with twin extruders. However, this technique uses fibers with very short length, i.e., particles/powder. The fibers and polymer are mixed by the twin extruder. The final product is in the form of pellets where they are used later in the formation of composites. Injection molding is used with thermoplastics and natural fibers to manufacture composites. One of the most important things in injection molding is that the polymer used should have low molecular weight. The low molecular weight means that the polymer will exhibit low viscosity; hence, processing will be easy. The difference between injection molding and extrusion is that in injection molding a permanent mold is used to produce the final part. Schematic diagrams of extrusion and injection molding are shown in Fig. 3.8a, b, respectively. A co-injection molding process was developed by Johnson Controls Automotive (JCA) to produce automotive interior parts from natural fibers and polyolefins named *COIXIL*. The final product of the COIXIL process is a sandwich structure. The feedstock of injection molding is usually in the form of short

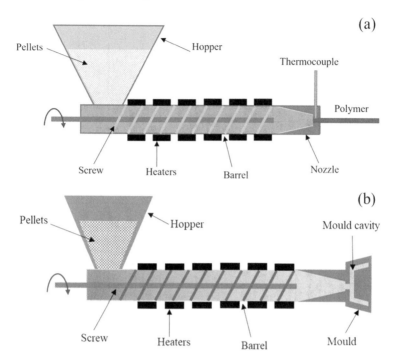

Fig. 3.8 Schematic diagram of **a** extrusion and **b** injection molding

fibers or particles. This is considered a disadvantage because the fibers in this case are regarded as inclusions that reduce the strength of the composites (Magurno 1999; Bledzki et al. 2006; Du et al. 2014). Several natural fiber composites were made in the literature using extrusion and injection molding processes (Abu-Sharkh and Hamid 2004; Bledzki et al. 2007; Alawar et al. 2008; Bledzki and Jaszkiewicz 2010; Dehghani et al. 2013; Mohanty 2017; Zadeh et al. 2017; Mazzanti et al. 2019; Kharrat et al. 2020; Malinowski et al. 2020).

Table 3.5 summarizes some of the attempts made in the literature to use natural fibers as reinforcement. The table compares between the work done in terms of fiber type and form, matrix type, and fabrication technique.

Table 3.5 Comparison between natural fiber-reinforced composites

Reinforcement	Fiber form	Polymer	Fabrication technique	References
Sisal + Glass	Long unidirectional	Epoxy	Film stacking/Compression molding	Prabhu et al. (2019)
Sisal	Woven, plain, twill, satin	Epoxy	Mixture on fibers/curing in closed mold	Srisuwan and Chumsamrong (2012)
Sisal	Hand weaving	Epoxy	Mixture on fibers/curing in closed mold	Srisuwan et al. (2014)
Sisal	Woven; 2 × 2	Natural rubber sheets	Compression molding	Jacob et al. (2006)
Sisal + Glass	Chopped 5 mm	Epoxy	Hand layup	Yuvaraj et al. (2017)
Sisal	Chopped 10 mm	Unsaturated polyester	Compression molding	Kejariwal and Keerthi Gowda (2017)
Sisal	Unidirectional	Unsaturated polyester	Compression molding	Senthilkumar et al. (2017)
Sisal	Chopped	*Self-reinforced*	Hot pressing	Lu et al. (2003)
Sisal + Cotton	Woven	Unsaturated polyester	Hand layup/compression molding	Sathishkumar et al. (2017a)
Sisal	Unidirectional	Unsaturated polyester/epoxy	Compression molding	Jiang et al. (2009)
Sisal + Kenaf	Woven, plain	Bio-epoxy	Compression molding	Yorseng et al. (2020)
Sisal	Mats	PP	Compression molding	Bajpai and Singh (2013)
Sisal + Jute	Woven; twill	Epoxy	Hand layup	de Araujo Alves Lima et al. (2019)
Sisal	Long nonwoven	Epoxy	Compression molding	Pushparaja et al. (2017)
Sisal	Nonwoven	PP	Compression molding	Prajwal et al. (2019)
Sisal	Chopped 50 mm	Perfluoroalkoxy alkene (PFA)	Mixture poured over fibers	Motaung et al. (2017)
Sisal + Glass	Woven	Epoxy	Hand layup	Palani Kumar et al. (2017)

(continued)

Table 3.5 (continued)

Reinforcement	Fiber form	Polymer	Fabrication technique	References
Sisal	Unidirectional	Epoxy	RTM	Li et al. (2015b)
Bamboo	Unidirectional	PLA	Compression molding	Roy Choudhury and Debnath (2020)
Bamboo	Emulsion	PLA	Extrusion/Compression molding	Li et al. (2019)
Bamboo	Unidirectional	Starch-based resin	Compression molding	Ochi (2006)
Coir	Powder	Biopolymer	Compression molding	Abdullah et al. (2020)
Coir	Chopped 10 mm	Unsaturated polyester	Compression molding	Keerthi Gowda et al. (2016)
Coir	Long nonwoven	Epoxy	Compression molding	Pushparaja et al. (2017)
Coir	Nonwoven	Epoxy	Hand layup/compression molding	Romli et al. (2012)
Coir	Short fibers	PP	Compression molding	Mir et al. (2013)
Coir	Nonwoven	Epoxy	Hand layup/compression molding	Suresh Kumar et al. (2014)
Coir	Ground	Unsaturated polyester	Compression molding	Mulinari et al. (2011)
Coir	Yarn	PLA	Commingling/compression molding	Jang et al. (2012)
Coir	Chopped	PP	Extrusion/injection molding	Haque et al. (2010a)
Coir	Chopped	PP	Compression molding	Lai et al. (2005)
Coir	Chopped	Epoxy	Hand layup/compression molding	Harish et al. (2009)
Coir + Sisal	Yarn	PP	Commingling/compression molding	Arya et al. (2015)

(continued)

Table 3.5 (continued)

Reinforcement	Fiber form	Polymer	Fabrication technique	References
Coir	Chopped	Polybutylene adipate terephthalate	Extrusion	Malinowski et al. (2020)
Kenaf + Oil palm	Nonwoven	Epoxy	Hand layup/compression molding	Hanan et al. (2018)
Kenaf + Banana	Woven; plain, twill/Nonwoven	Unsaturated polyester	Hand layup	Alavudeen et al. (2015)
Kenaf	Needle punch	Starch glue	Hot pressing	Husain et al. (2019)
Kenaf + Glass	Mats	Epoxy	Modified SMC	Davoodi et al. (2010)
Kenaf	Twisted fibers	PP	Hot impregnation	Jeyanthi and Rani (2012)
Kenaf	Chopped	HDPE	Extrusion	Salleh et al. (2014)
Kenaf	Chopped	HDPE/PP	Hot press machine	Meon et al. (2012)
Kenaf	Nonwoven	PP	Compression molding	Asumani et al. (2012)
Kenaf	Chopped	PP	Compression molding	Zampaloni et al. (2007)
Kenaf	Mats	PP	Carding/hot pressing	Islam et al. (2011)
Kenaf + Bamboo + Jute	Mats	Unsaturated polyester	Hand layup/compression molding	Hojo et al. (2014)
Kenaf	Chopped	PP	Hot pressing	Shibata et al. (2006)
Kenaf	Chopped	PLA	Carding/needle punching/hot pressing	Lee et al. (2009)
Kenaf	Chopped	PP	Extrusion/compression molding	Bernard et al. (2011)

(continued)

Table 3.5 (continued)

Reinforcement	Fiber form	Polymer	Fabrication technique	References
Kenaf	Chopped	PP	Carding/needle punching/compression molding	Hao et al. (2013)
Kenaf	Unidirectional	Epoxy	Hand layup	Yousif et al. (2012)
Kenaf	Nonwoven	Kevlar/Epoxy	Hand layup	Yahaya et al. (2014)
Kenaf	Nonwoven	Epoxy	Hand layup	Alkbir et al. (2016)
Hemp	Needle punch	PLA	Hot pressing	Xu et al. (2019)
Hemp	Woven	Vinyl ester	Hand layup	Misnon et al. (2016)
Hemp	Needle punching	PP		Yilmaz et al. (2012)
Jute	Woven; plain	PLA	Film stacking/compression molding	Papa et al. (2017)
Jute	Woven; plain/unidirectional	PLLA	Compression molding	Khan et al. (2016)
Jute	Unidirectional	Starch-based resin	Compression molding	Ochi (2006)
Jute	Woven; plain	Unsaturated polyester	VARTM	Santulli (2000)
Jute + Flax + Basalt	Woven	Epoxy	Vacuum bag	Oliver-Borrachero et al. (2019)
Abaca + Hemp + Flax	Chopped 25 mm	PP	Injection molding	Bledzki et al. (2007)
Abaca	Unidirectional	Starch-based resin	Compression molding	Ochi (2006)
Abaca	Short fibers	PP	Injection molding	Bledzki et al. (2010)
Ramie	Unidirectional	Starch-based resin	Compression molding	Ochi (2006)
Ramie	Powder	HDPE powder	Compression molding	Banowati et al. (2016)
Flax	Woven; hopsack, twill/unidirectional	Epoxy	VARTM	Rayyaan et al. (2019)

(continued)

Table 3.5 (continued)

Reinforcement	Fiber form	Polymer	Fabrication technique	References
Flax + Glass	Woven; twill	Epoxy	Autoclave	Asgarinia et al. (2015)
Flax + Hemp	Chopped	Soybean oil	VARTM	O'Donnell et al. (2004)
Flax	Needle punch	Low melt PET	Hot calendaring machine	Muthukumar et al. (2019)
Flax	Prepreg – Quasi Unidirectional	Epoxy	Hand layup	Li et al. (2015a)
Flax	Lineo Unidirectional	Epoxy	Compression molding	Berges et al. (2016)
Flax	DREF-3	PP	Commingling/Compression molding	Bar et al. (2018)
Flax	Woven; 4 × 4 hopsack	Bioplast films	Compression molding	Duquesne et al. (2015)
Flax	Woven; 4 × 4 plain, 2 × 2 twill	Polybutylene adipate terephthalate	Compression molding	Phongam et al. (2015)
Flax + Kenaf + Glass	Nonwoven	Epoxy	Hand layup	Srinivasan et al. (2014)
Flax	Woven; knitted	Epoxy	Hand layup	Muralidhar et al. (2012)
Jute + Flax + Glass	Randomly oriented	Epoxy	Hand layup	Ramnath et al. (2014)
Flax	Fabric form	Epoxy	VARTM	Bertomeu et al. (2012)
Flax	Woven; plain	Epoxy	Hand layup	Muralidhar (2013)
Flax	Biaxial	Epoxy	Hand layup/vacuum bagging	Di Bella et al. (2010)
Flax + Jute	Woven; plain	Epoxy	Hand layup/vacuum bagging	Dhakal et al. (2014)
Flax	Unidirectional	Unsaturated polyester/ethylene propylene	Hand layup/RTM	Goutianos et al. (2007)

References

Abdal-hay A, Suardana NPG, Jung DY et al (2012) Effect of diameters and alkali treatment on the tensile properties of date palm fiber reinforced epoxy composites. Int J Precis Eng Manuf 13:1199–1206. https://doi.org/10.1007/s12541-012-0159-3

Abdullah M, Alharbi H, Hirai S et al (2020) Effects of chemical composition, mild alkaline pretreatment and particle size on mechanical, thermal, and structural properties of binderless lignocellulosic biopolymers prepared by hot-pressing raw microfibrillated Phoenix dactylifera and Cocos nucifera. Polym Test 106384. https://doi.org/10.1016/j.polymertesting.2020.106384

Abu-Sharkh BF, Hamid H (2004) Degradation study of date palm fibre/polypropylene composites in natural and artificial weathering: mechanical and thermal analysis. Polym Degrad Stab 85:967–973. https://doi.org/10.1016/j.polymdegradstab.2003.10.022

Al-Kaabi K, Al-Khanbashi A, Hammami A (2005) Date palm fibers as polymeric matrix reinforcement: DPF/polyester composite properties. Polym Compos 26:604–613. https://doi.org/10.1002/pc.20130

Alajmi M, Shalwan A (2015) Correlation between mechanical properties with specific wear rate and the coefficient of friction of graphite/epoxy composites. Materials (basel) 8:4162–4175. https://doi.org/10.3390/ma8074162

Alavudeen A, Rajini N, Karthikeyan S et al (2015) Mechanical properties of banana/kenaf fiber-reinforced hybrid polyester composites: effect of woven fabric and random orientation. Mater Des 66:246–257. https://doi.org/10.1016/j.matdes.2014.10.067

Alawar A, Hamed AM, Al-Kaabi K (2008) Date palm tree fiber as polymeric matrix reinforcement, DPF-polypropylene composite characterization. Adv Mater Res 47–50:193–196. https://doi.org/10.4028/www.scientific.net/AMR.47-50.193

Ali ME, Alabdulkarem A (2017) On thermal characteristics and microstructure of a new insulation material extracted from date palm trees surface fibers. Constr Build Mater 138:276–284. https://doi.org/10.1016/j.conbuildmat.2017.02.012

Alkbir MFM, Sapuan SM, Nuraini AA, Ishak MR (2016) The effect of fiber content on the crashworthiness parameters of natural kenaf fiber-reinforced hexagonal composite tubes. J Eng Fiber Fabr 11:75–86. https://doi.org/10.1177/155892501601100102

Anidha S, Latha N, Muthukkumar M (2020) Effect of polyaramid reinforced with sisal epoxy composites: tensile, impact, flexural and morphological properties. J Mater Res Technol 9:7947–7954. https://doi.org/10.1016/j.jmrt.2020.04.081

Arya A, Tomlal JE, Gejo G, Kuruvilla J (2015) Commingled composites of polypropylene/coir-sisal yarn: Effect of chemical treatments on thermal and tensile properties. E-Polymers 15:169–177. https://doi.org/10.1515/epoly-2014-0186

Asgarinia S, Viriyasuthee C, Phillips S et al (2015) Tension-tension fatigue behaviour of woven flax/epoxy composites. J Reinf Plast Compos 34:857–867. https://doi.org/10.1177/0731684415581527

Asumani OML, Reid RG, Paskaramoorthy R (2012) The effects of alkali-silane treatment on the tensile and flexural properties of short fibre non-woven kenaf reinforced polypropylene composites. Compos Part A Appl Sci Manuf 43:1431–1440. https://doi.org/10.1016/j.compositesa.2012.04.007

Azammi AMN, Sapuan SM, Ishak MR, Sultan MTH (2020) Physical and damping properties of kenaf fibre filled natural rubber/thermoplastic polyurethane composites. Def Technol 16:29–34. https://doi.org/10.1016/j.dt.2019.06.004

Bajpai PK, Singh I (2013) Drilling behavior of sisal fiber-reinforced polypropylene composite laminates. J Reinf Plast Compos 32:1569–1576. https://doi.org/10.1177/0731684413492866

Banowati L, Hadi BK, Suratman R, Faza A (2016) Tensile strength of ramie yarn (spinning by machine)/HDPE thermoplastic matrix composites. AIP Conf Proc 1717:4. https://doi.org/10.1063/1.4943456

Bar M, Alagirusamy R, Das A (2019) Development of flax-PP based twist-less thermally bonded roving for thermoplastic composite reinforcement. J Text Inst 110:1369–1379. https://doi.org/10. 1080/00405000.2019.1610997

Bar M, Das A, Alagirusamy R (2018) Effect of interface on composites made from DREF spun hybrid yarn with low twisted core flax yarn. Compos Part A Appl Sci Manuf 107:260–270. https:// doi.org/10.1016/j.compositesa.2018.01.003

Berges M, Léger R, Placet V et al (2016) Influence of moisture uptake on the static, cyclic and dynamic behaviour of unidirectional flax fibre-reinforced epoxy laminates. Compos Part A Appl Sci Manuf 88:165–177. 10.1016/j.compositesa.2016.05.029

Bernard M, Khalina A, Ali A et al (2011) The effect of processing parameters on the mechanical properties of kenaf fibre plastic composite. Mater Des 32:1039–1043. 10.1016/j.matdes.2010.07.014

Bertomeu D, Garcı́a-Sanoguera D, Fenollar O, et al (2012) Use of eco-friendly epoxy resins from renewable resources as potential substitutes of petrochemical epoxy resins for ambient cured composites with flax reinforcements. Polym Compos 33:683–692. https://doi.org/10.1002/pc

Bledzki AK, Faruk O, Sperber VE (2006) Cars from Bio-Fibres. Macromol Mater Eng 291:449–457. https://doi.org/10.1002/mame.200600113

Bledzki AK, Jaszkiewicz A (2010) Mechanical performance of biocomposites based on PLA and PHBV reinforced with natural fibres—A comparative study to pp. Compos Sci Technol 70:1687– 1696. https://doi.org/10.1016/j.compscitech.2010.06.005

Bledzki AK, Mamun AA, Faruk O (2007) Abaca fibre reinforced PP composites and comparison with jute and flax fibre PP composites. Express Polym Lett 1:755–762. https://doi.org/10.3144/ expresspolymlett.2007.104

Bledzki AK, Mamun AA, Jaszkiewicz A, Erdmann K (2010) Polypropylene composites with enzyme modified abaca fibre. Compos Sci Technol 70:854–860. https://doi.org/10.1016/j.com pscitech.2010.02.003

Bledzki AK, Sperber VE, Faruk O (2002) Natural and wood fibre reinforcement in polymers. Macromolecules 13:3–144

Boutin M, Rogeon A, Aufray M et al (2020) Influence of flax fibers on network formation of DGEBA/DETA matrix. Compos Interfaces 00:1–18. https://doi.org/10.1080/09276440.2020.173 6454

Davoodi MM, Sapuan SM, Ahmad D et al (2010) Mechanical properties of hybrid kenaf/glass reinforced epoxy composite for passenger car bumper beam. Mater Des 31:4927–4932. https:// doi.org/10.1016/j.matdes.2010.05.021

de Araujo Alves Lima R, Kawasaki Cavalcanti D, de Souza e Silva Neto J et al (2019) Effect of surface treatments on interfacial properties of natural intralaminar hybrid composites. Polym Compos 314–325. https://doi.org/10.1002/pc.25371

Dehghani A, Madadi Ardekani S, Al-Maadeed MA et al (2013) Mechanical and thermal properties of date palm leaf fiber reinforced recycled poly (ethylene terephthalate) composites. Mater Des 52:841–848. https://doi.org/10.1016/j.matdes.2013.06.022

del Borrello M, Mele M, Campana G, Secchi M (2020) Manufacturing and characterization of hemp-reinforced epoxy composites. Polym Compos 41:2316–2329. https://doi.org/10.1002/pc. 25540

Dhakal HN, Zhang ZY, Bennett N et al (2014) Effects of water immersion ageing on the mechanical properties of flax and jute fibre biocomposites evaluated by nanoindentation and flexural testing. J Compos Mater 48:1399–1406. https://doi.org/10.1177/0021998313487238

Di Bella G, Fiore V, Valenza A (2010) Effect of areal weight and chemical treatment on the mechanical properties of bidirectional flax fabrics reinforced composites. Mater Des 31:4098–4103. https://doi.org/10.1016/j.matdes.2010.04.050

Du Y, Yan N, Kortschot MT (2014) A simplified fabrication process for biofiber-reinforced polymer composites for automotive interior trim applications. J Mater Sci 49:2630–2639. https://doi.org/ 10.1007/s10853-013-7965-6

Duquesne S, Samyn F, Ouagne P, Bourbigot S (2015) Flame retardancy and mechanical properties of flax reinforced woven for composite applications. J Ind Text 44:665–681. https://doi.org/10.1177/1528083713505633

Elseify LA, Midani M (2020) Characterization of date palm fiber. In: Midani M, Saba N, Alothman OY (eds) Date palm fiber composites, 1st edn. Springer, Singapore, pp 227–255

Goutianos S, Peijs T, Nystrom B, Skrifvars M (2007) Textile reinforcements based on aligned flax fibres for structural composites. Compos Innov 1–13

Hanan F, Jawaid M, Md Tahir P (2018) Mechanical performance of oil palm/kenaf fiber-reinforced epoxy-based bilayer hybrid composites. J Nat Fibers 00:1–13. https://doi.org/10.1080/15440478.2018.1477083

Hao A, Zhao H, Chen JY (2013) Kenaf/polypropylene nonwoven composites: The influence of manufacturing conditions on mechanical, thermal, and acoustical performance. Compos Part B Eng 54:44–51. https://doi.org/10.1016/j.compositesb.2013.04.065

Haque M, Islam N, Huque M et al (2010a) Coir fiber reinforced polypropylene composites: Physical and mechanical properties. Adv Compos Mater 19:91–106. https://doi.org/10.1163/092430409X12530067339325

Haque M, Islam S, Islam S et al (2010b) Physicomechanical properties of chemically treated palm fiber reinforced polypropylene composites. J Reinf Plast Compos 29:1734–1742. https://doi.org/10.1177/0731684409341678

Harish S, Michael DP, Bensely A et al (2009) Mechanical property evaluation of natural fiber coir composite. Mater Charact 60:44–49. https://doi.org/10.1016/j.matchar.2008.07.001

Hassan T, Jamshaid H, Mishra R et al (2020) Acoustic, mechanical and thermal properties of green composites reinforced with natural fiberswaste. Polymers (Basel) 12. https://doi.org/10.3390/polym12030654

Hassanin AH, Elseify LA, Hamouda T (2020) Date palm fiber composite fabrication techniques. In: Midani M, Saba N, Alothman OY (eds) Date palm fiber composites, 1st edn. Springer, Singapore, pp 161–183

Hojo T, Zhilan XU, Yang Y, Hamada H (2014) Tensile properties of bamboo, jute and kenaf mat-reinforced composite. Energy Procedia 56:72–79. https://doi.org/10.1016/j.egypro.2014.07.133

Husain SNH, Abdul Razak NA, Abdul Rashid AH et al (2019) Development of kenaf nonwoven as automotive noise absorption. Appl Mech Mater 892:101–105. https://doi.org/10.4028/www.scientific.net/amm.892.101

Ibrahim H, Farag M, Megahed H, Mehanny S (2014) Characteristics of starch-based biodegradable composites reinforced with date palm and flax fibers. Carbohydr Polym 101:11–19. https://doi.org/10.1016/j.carbpol.2013.08.051

Ibrahim H, Mehanny S, Darwish L, Farag M (2017) A comparative study on the mechanical and biodegradation characteristics of starch-based composites reinforced with different lignocellulosic fibers. J Polym Environ 0:1–14. https://doi.org/10.1007/s10924-017-1143-x

Islam MS, Church JS, Miao M (2011) Effect of removing polypropylene fibre surface finishes on mechanical performance of kenaf/polypropylene composites. Compos Part A Appl Sci Manuf 42:1687–1693. https://doi.org/10.1016/j.compositesa.2011.07.023

Jaafar J, Siregar JP, Tezara C et al (2019) A review of important considerations in the compression molding process of short natural fiber composites. Int J Adv Manuf Technol 3437–3450. https://doi.org/10.1007/s00170-019-04466-8

Jacob M, Varughese KT, Thomas S (2006) A study on the moisture sorption characteristics in woven sisal fabric reinforced natural rubber biocomposites. J Appl Polym Sci 102:416–423. https://doi.org/10.1002/app.24061

Jang JY, Jeong TK, Oh HJ et al (2012) Thermal stability and flammability of coconut fiber reinforced poly(lactic acid) composites. Compos Part B Eng 43:2434–2438. https://doi.org/10.1016/j.compositesb.2011.11.003

Jeyanthi S, Rani JJ (2012) Influence of natural long fiber in mechanical, thermal and recycling properties of thermoplastic composites in automotive components. Int J Phys Sci 7:5765–5771. https://doi.org/10.5897/IJPS12.521

Jiang X, Rui Y, Chen G (2009) Unsaturated polyester-toughened epoxy composites: effect of sisal fiber on thermal and dynamic mechanical properties of sisal fiber on thermal and dynamic mechanical properties. J Vinyl Addit Technol 21:129–133. https://doi.org/10.1002/vnl

Keerthi Gowda BS, Easwara Prasad GL, Velmurgan, (2016) Probabilistic study of tensile properties of coir fiber reinforced polymer matrix composite probabilistic study of tensile properties of coir fiber reinforced polymer matrix composite. Int J Adv Mater Sci 06:7–17

Kejariwal RK, Keerthi Gowda BS (2017) Flammability and moisture absorption behavior of sisal-polyester composites. Mater Today Proc 4:8040–8044. https://doi.org/10.1016/j.matpr.2017.07.142

Khan GMA, Terano M, Gafur MA, Alam MS (2016) Studies on the mechanical properties of woven jute fabric reinforced poly(L-lactic acid) composites. J King Saud Univ Eng Sci 28:69–74. https://doi.org/10.1016/j.jksues.2013.12.002

Kharrat F, Khlif M, Hilliou L, et al (2020) Minimally processed date palm (*Phoenix dactylifera* L.) leaves as natural fillers and processing aids in poly(lactic acid) composites designed for the extrusion film blowing of thin packages. Ind Crops Prod 154:112637. https://doi.org/10.1016/j.indcrop.2020.112637

Lai CY, Sapuan SM, Ahmad M et al (2005) Mechanical and electrical properties of coconut coir fiber-reinforced polypropylene composites. Polym Plast Technol Eng 44:619–632. https://doi.org/10.1081/PTE-200057787

Lee BH, Kim HS, Lee S et al (2009) Bio-composites of kenaf fibers in polylactide: role of improved interfacial adhesion in the carding process. Compos Sci Technol 69:2573–2579. https://doi.org/10.1016/j.compscitech.2009.07.015

Li W, He X, Zuo Y et al (2019) Study on the compatible interface of bamboo fiber/polylactic acid composites by in-situ solid phase grafting. Int J Biol Macromol 141:325–332. https://doi.org/10.1016/j.ijbiomac.2019.09.005

Li Y, Li Q, Ma H (2015a) The voids formation mechanisms and their effects on the mechanical properties of flax fiber reinforced epoxy composites. Compos Part A Appl Sci Manuf 72:40–48. https://doi.org/10.1016/j.compositesa.2015.01.029

Li Y, Ma H, Shen Y et al (2015b) Effects of resin inside fiber lumen on the mechanical properties of sisal fiber reinforced composites. Compos Sci Technol 108:32–40. https://doi.org/10.1016/j.compscitech.2015.01.003

Lu X, Zhang MQ, Rong MZ et al (2003) Self-reinforced melt processable composites of sisal. Compos Sci Technol 63:177–186. https://doi.org/10.1016/S0266-3538(02)00204-X

Magurno A (1999) Vegetable fibres in automotive interior components. Angew Makromol Chemie 272:99–107. https://doi.org/10.1002/(SICI)1522-9505(19991201)272:1%3c99::AID-APMC99%3e3.0.CO;2-C

Mahdi E, Hernández DR, Eltai EO (2015) Effect of water absorption on the mechanical properties of long date palm leaf fiber reinforced epoxy composites. J Biobased Mater Bioenergy 9:173–181

Malinowski R, Krasowska K, Sikorska W, et al (2020) Studies on manufacturing, mechanical properties and structure of poly(butylene adipate-co-terephthalate)-based green composites modified by coconut fibers. Int J Precis Eng Manuf Green Technol. https://doi.org/10.1007/s40684-019-00171-9

Masri T, Ounis H, Sedira L et al (2018) Characterization of new composite material based on date palm leaflets and expanded polystyrene wastes. Constr Build Mater 164:410–418. https://doi.org/10.1016/j.conbuildmat.2017.12.197

Mazzanti V, Pariante R, Bonanno A et al (2019) Reinforcing mechanisms of natural fibers in green composites: role of fibers morphology in a PLA/hemp model system. Compos Sci Technol 180:51–59. https://doi.org/10.1016/j.compscitech.2019.05.015

Mekonnen BY, Mamo YJ (2020) Tensile and flexural analysis of a hybrid bamboo/jute fiber-reinforced composite with polyester matrix as a sustainable green material for wind turbine blades. Int J Eng Trans B Appl 33:314–319. https://doi.org/10.5829/IJE.2020.33.02B.16

Meon MS, Othman MF, Husain H et al (2012) Improving tensile properties of kenaf fibers treated with sodium hydroxide. Procedia Eng 41:1587–1592. https://doi.org/10.1016/j.proeng. 2012.07.354

Mir SS, Nafsin N, Hasan M et al (2013) Improvement of physico-mechanical properties of coir-polypropylene biocomposites by fiber chemical treatment. Mater Des 52:251–257. https://doi. org/10.1016/j.matdes.2013.05.062

Mirmehdi SM, Zeinaly F, Dabbagh F (2014) Date palm wood flour as filler of linear low-density polyethylene. Compos Part B Eng 56:137–141. https://doi.org/10.1016/j.compositesb. 2013.08.008

Mishra R, Wiener J, Militky J et al (2020) Bio-composites reinforced with natural fibers: comparative analysis of thermal, static and dynamic-mechanical properties. Fibers Polym 21:619–627. https:// doi.org/10.1007/s12221-020-9804-0

Misnon MI, Islam MM, Epaarachchi A (2016) Fabric Parameter effect on the mechanical properties of woven hemp fabric reinforced composites as an alternative to wood product. Adv Res Text Eng 1:1004

Mohanty JR (2017) Investigation on solid particle erosion behavior of date palm leaf fiber-reinforced polyvinyl pyrrolidone composites. J Thermoplast Compos Mater 30:1003–1016. https://doi.org/ 10.1177/0892705715614079

Mohanty JR, Das SN, Das HC, Swain SK (2014) Effect of chemically modified date palm leaf fiber on mechanical, thermal and rheological properties of polyvinylpyrrolidone. Fibers Polym 15:1062–1070. https://doi.org/10.1007/s12221-014-1062-6

Motaung TE, Linganiso LZ, Kumar R, Anandjiwala RD (2017) Agave and sisal fibre-reinforced polyfurfuryl alcohol composites. J Thermoplast Compos Mater 30:1323–1343. https://doi.org/ 10.1177/0892705716632858

Mulinari DR, Baptista CARP, Souza JVC, Voorwald HJC (2011) Mechanical properties of coconut fibers reinforced polyester composites. Procedia Eng 10:2074–2079. https://doi.org/10.1016/j. proeng.2011.04.343

Muralidhar BA (2013) Tensile and compressive properties of flax-plain weave preform reinforced epoxy composites. J Reinf Plast Compos 32:207–213. https://doi.org/10.1177/073168441246 9136

Muralidhar BA, Giridev VR, Raghunathan K (2012) Flexural and impact properties of flax woven, knitted and sequentially stacked knitted/woven preform reinforced epoxy composites. J Reinf Plast Compos 31:379–388. https://doi.org/10.1177/0731684412437987

Muthukumar N, Thilagavathi G, Neelakrishnan S, Poovaragan PT (2019) Sound and thermal insulation properties of flax/low melt PET needle punched nonwovens. J Nat Fibers 16:245–252. https://doi.org/10.1080/15440478.2017.1414654

Neher B, Bhuiyan MMR, Kabir H et al (2016) Thermal properties of palm fiber and palm fiber-reinforced ABS composite. J Therm Anal Calorim 124:1281–1289. https://doi.org/10.1007/s10 973-016-5341-x

O'Donnell A, Dweib MA, Wool RP (2004) Natural fiber composites with plant oil-based resin. Compos Sci Technol 64:1135–1145. https://doi.org/10.1016/j.compscitech.2003.09.024

Ochi S (2006) Development of high strength biodegradable composites using Manila hemp fiber and starch-based biodegradable resin. Compos Part A Appl Sci Manuf 37:1879–1883. https:// doi.org/10.1016/j.compositesa.2005.12.019

Oliver-Borrachero B, Sánchez-Caballero S, Fenollar O, Sellés MA (2019) Natural-Fiber-Reinforced Polymer Composites for Automotive Parts Manufacturing. Key Eng Mater 793:9–16. https://doi. org/10.4028/www.scientific.net/kem.793.9

Ouarhim W, Essabir H, Bensalah MO et al (2020) Hybrid composites and intra-ply hybrid composites based on jute and glass fibers: A comparative study on moisture absorption and mechanical properties. Mater Today Commun 22. https://doi.org/10.1016/j.mtcomm.2019.100861

Ovlaque P, Bayart M, Cousin P et al (2020) Mechanical & Interfacial properties of bamboo lamella-PP composites—Effect of lamella treatment. Fibers Polym 21:1086–1095. https://doi.org/10. 1007/s12221-020-9526-3

Palani Kumar K, Shadrach Jeya Sekaran A, Pitchandi K (2017) Investigation on mechanical properties of woven alovera/sisal/kenaf fibres and their hybrid composites. Bull Mater Sci 40:117–128.
https://doi.org/10.1007/s12034-016-1343-3

Papa I, Lopresto V, Simeoli G et al (2017) Ultrasonic damage investigation on woven jute/poly
(lactic acid) composites subjected to low velocity impact. Compos Part B Eng 115:282–288.
https://doi.org/10.1016/j.compositesb.2016.09.076

Phongam N, Dangtungee R, Siengchin S (2015) Comparative studies on the mechanical properties of nonwoven- and woven-flax-fiber-reinforced poly(butylene adipate-co-terephthalate)-
based composite laminates. Mech Compos Mater 51:17–24. https://doi.org/10.1007/s11029-015-
9472-0

Prabhu L, Krishnaraj V, Gokulkumar S et al (2019) Mechanical, chemical and acoustical behavior of
sisal—Tea waste—Glass fiber reinforced epoxy based hybrid polymer composites. Mater Today
Proc 16:653–660. https://doi.org/10.1016/j.matpr.2019.05.142

Prajwal B, Giridharan BV, Vamshi KK et al (2019) Sisal fiber reinforced polypropylene biocomposites for inherent applications. Int J Recent Technol Eng 8:305–309

Pushparaja, Balaganesan G, Velmurugan R (2017) Frangibility study of natural fiber reinforced
composite laminates. Key Eng Mater 725 KEM:88–93. https://doi.org/10.4028/www.scientific.
net/KEM.725.88

Ramnath BV, Elanchezhian C, Nirmal PV et al (2014) Experimental investigation of mechanical
behavior of Jute-Flax based glass fiber reinforced composite. Fibers Polym 15:1251–1262. https://
doi.org/10.1007/s12221-014-1251-3

Rayyaan R, Kennon WR, Potluri P, Akonda M (2019) Fibre architecture modification to improve
the tensile properties of flax-reinforced composites. J Compos Mater 002199831986315. https://
doi.org/10.1177/0021998319863156

Rohen LA, Margem FM, Monteiro SN et al (2015) Ballistic efficiency of an individual epoxy
composite reinforced with sisal fibers in multilayered armor. Mater Res 18:55–62. https://doi.
org/10.1590/1516-1439.346314

Rokbi M, Khaldoune A, Sanjay MR et al (2020) Effect of processing parameters on tensile properties
of recycled polypropylene based composites reinforced with jute fabrics. Int J Light Mater Manuf
3:144–149. https://doi.org/10.1016/j.ijlmm.2019.09.005

Romli FI, Alias AN, Rafie ASM, Majid DLAA (2012) Factorial study on the tensile strength of
a coir fiber-reinforced epoxy composite. AASRI Procedia 3:242–247. https://doi.org/10.1016/j.
aasri.2012.11.040

Roy Choudhury M, Debnath K (2020) A study of drilling behavior of unidirectional bamboo
fiber-reinforced green composites. J Inst Eng Ser C. https://doi.org/10.1007/s40032-019-00550-w

Saleh MA, Al Haron MH, Saleh AA, Farag M (2017) Fatigue behavior and life prediction of
biodegradable composites of starch reinforced with date palm fibers. Int J Fatigue 103:216–222.
https://doi.org/10.1016/j.ijfatigue.2017.06.005

Salleh FM, Hassan A, Yahya R, Azzahari AD (2014) Effects of extrusion temperature on the
rheological, dynamic mechanical and tensile properties of kenaf fiber/HDPE composites. Compos
Part B Eng 58:259–266. https://doi.org/10.1016/j.compositesb.2013.10.068

Santulli C (2000) Mechanical and impact properties of untreated jute fabric reinforced polyester
laminates compared with different E-glass fibre reinforced laminates. Sci Eng Compos Mater
9:177–188. https://doi.org/10.1515/secm.2000.9.4.177

Sathishkumar S, Suresh AV, Nagamadhu M, Krishna M (2017a) The effect of alkaline treatment on
their properties of Jute fiber mat and its vinyl ester composites. Mater Today Proc 4:3371–3379.
https://doi.org/10.1016/j.matpr.2017.02.225

Sathishkumar TP, Naveen J, Navaneethakrishnan P et al (2017b) Characterization of sisal/cotton
fibre woven mat reinforced polymer hybrid composites. J Ind Text 47:429–452. https://doi.org/
10.1177/1528083716648764

Sbiai A, Kaddami H, Fleury E et al (2008) Effect of the fiber size on the physicochemical and
mechanical properties of composites of epoxy and date palm tree fibers. Macromol Mater Eng
293:684–691. https://doi.org/10.1002/mame.200800087

Senthilkumar K, Siva I, Sultan MTH et al (2017) Static and dynamic properties of sisal fiber polyester composites—Effect of interlaminar fiber orientation. BioResources 12:7819–7833. https://doi.org/10.15376/biores.12.4.7819-7833

Shalwan A, Yousif BF (2014) Influence of date palm fibre and graphite filler on mechanical and wear characteristics of epoxy composites. Mater Des 59:264–273. https://doi.org/10.1016/j.matdes.2014.02.066

Shibata S, Cao Y, Fukumoto I (2006) Lightweight laminate composites made from kenaf and polypropylene fibres. Polym Test 25:142–148. https://doi.org/10.1016/j.polymertesting.2005.11.007

Srinivasan VS, Rajendra Boopathy S, Vijaya Ramnath B (2014) Investigation on shear behaviour of flax-kenaf hybrid composite. Adv Mater Res 1051:139–142. https://doi.org/10.4028/www.scientific.net/AMR.1051.139

Srisuwan S, Chumsamrong P (2012) The effects of fiber architecture and fiber surface treatment on physical properties of woven sisal fiber/epoxy composites. Adv Mater Res 410:36–42. https://doi.org/10.4028/www.scientific.net/AMR.410.39

Srisuwan S, Prasoetsopha N, Suppakarn N, Chumsamrong P (2014) The effects of alkalized and silanized woven sisal fibers on mechanical properties of natural rubber modified epoxy resin. Energy Procedia 56:19–25. https://doi.org/10.1016/j.egypro.2014.07.127

Summerscales J, Grove S (2013) Manufacturing methods for natural fibre composites. Woodhead Publishing Limited

Suresh Kumar SM, Duraibabu D, Subramanian K (2014) Studies on mechanical, thermal and dynamic mechanical properties of untreated (raw) and treated coconut sheath fiber reinforced epoxy composites. Mater Des 59:63–69. https://doi.org/10.1016/j.matdes.2014.02.013

Syaqira SN, Leman Z, Sapuan SM et al (2020) Tensile strength and moisture absorption of sugar palm-polyvinyl butyral laminated composites. Polymers (Basel) 12. https://doi.org/10.3390/POLYM12091923

Tataru G, Guibert K, Labbé M, et al (2020) Modification of flax fiber fabrics by radiation grafting: Application to epoxy thermosets and potentialities for silicone-natural fibers composites. Radiat Phys Chem 170:108663. https://doi.org/10.1016/j.radphyschem.2019.108663

Wazzan AA (2006) Effect of fiber orientation on the mechanical properties and fracture characteristics of date palm fiber reinforced composites. 213–225

Xu Z, Yang L, Ni Q, et al (2019) Fabrication of high-performance green hemp/polylactic acid fibre composites. J Eng Fiber Fabr 14. https://doi.org/10.1177/1558925019834497

Yahaya R, Sapuan SM, Jawaid M et al (2014) Quasi-static penetration and ballistic properties of kenaf-aramid hybrid composites. Mater Des 63:775–782. https://doi.org/10.1016/j.matdes.2014.07.010

Yilmaz ND, Powell NB, Banks-Lee P, Michielsen S (2012) Hemp-fiber based nonwoven composites: Effects of alkalization on sound absorption performance. Fibers Polym 13:915–922. https://doi.org/10.1007/s12221-012-0915-0

Yorseng K, Rangappa SM, Pulikkalparambil H et al (2020) Accelerated weathering studies of kenaf/sisal fiber fabric reinforced fully biobased hybrid bioepoxy composites for semi-structural applications: Morphology, thermo-mechanical, water absorption behavior and surface hydrophobicity. Constr Build Mater 235:117464. https://doi.org/10.1016/j.conbuildmat.2019.117464

Yousif BF, Shalwan A, Chin CW, Ming KC (2012) Flexural properties of treated and untreated kenaf/epoxy composites. Mater Des 40:378–385. https://doi.org/10.1016/j.matdes.2012.04.017

Yuvaraj G, Kumar H, Saravanan G (2017) An experimentation of chemical and mechanical behaviour of epoxy-sisal reinforced composites. Polym Polym Compos 25:221–224. https://doi.org/10.1177/096739111702500307

Zadeh KM, Inuwa IM, Arjmandi R et al (2017) Effects of date palm leaf fiber on the thermal and tensile properties of recycled ternary polyolefin blend composites. Fibers Polym 18:1330–1335. https://doi.org/10.1007/s12221-017-1106-9

Zampaloni M, Pourboghrat F, Yankovich SA et al (2007) Kenaf natural fiber reinforced polypropy-lene composites: A discussion on manufacturing problems and solutions. Compos Part A Appl Sci Manuf 38:1569–1580. https://doi.org/10.1016/j.compositesa.2007.01.001

Chapter 4
Natural Fiber Composite Qualification in the Automotive Industry

Abstract Materials qualification is an essential process for approving any new material by the automotive industry. The process occurs between material suppliers and automakers to ensure that any new material meets the performance requirements. This chapter is dedicated to discussing the natural fiber composite (NFC) qualification process in the automotive industry. The three-step process for introducing new materials to the automotive industry is explained. The major challenges facing the use of NFC in the industry are also discussed, including fiber availability, variability of properties, hydrophilicity, compatibility with resin, and degradability. The common automotive interior parts that are made from NFC are tabulated, with details on fiber and matrix types and major performance requirements, this includes door panels, front and rear door liners, headliners, rear parcel shelves, seat backs, spare tire covers, and other interior trims.

Keywords Materials qualifications · Door panel · Natural fiber composite

4.1 Materials Qualifications and Challenges

Materials qualification is a process that occurs between material suppliers and automakers for tier 1 supplier. The main aim of this process is to make sure that the materials properties meet the performance requirements. Materials qualification process follows a series of steps until the final products/properties are approved by the customer. It is preceded by materials development process and followed by validation as shown in Fig. 4.1. This step is essential and required because it is difficult for customers to use a newly developed material without making sure first that it fits the requirements especially in automotive applications (Shashank 2016).

In the material qualification process, the new material is subjected to several standardized tests for full characterization. The materials that will be used in structural applications are inspected and analyzed more than that used in non-structural applications. As a matter of fact, automakers are always reluctant, as mentioned earlier, to use a newly developed material since it was not tried before and many data are still not available (Shashank 2016). A very interesting article was written by Midani et al.

© The Author(s), under exclusive license to Springer Nature Switzerland AG 2021
L. A. Elseify et al., *Manufacturing Automotive Components from Sustainable Natural Fiber Composites*, SpringerBriefs in Materials,
https://doi.org/10.1007/978-3-030-83025-0_4

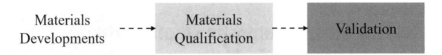

Fig. 4.1 Three-step process for introducing a new material to the automotive industry

(2019) studying and discussing why natural fibers are still not utilized as expected and what really hinders their usage in a larger scale. As mentioned in that article, the diffusion of innovation theory, by E. M. Rogers, could show why natural fibers are still not considered widely in mass production. The theory explains generally how new ideas and technologies are perceived by consumers. As shown in Fig. 4.2, there are 2 types of consumers: visionaries and pragmatists. Visionaries always tend to accept new ideas and innovations. On the other hand, pragmatists are the complete opposite; they tend to resist trying new things. And it takes years for a new idea to get accepted and adopted (Midani et al. 2019). As can be seen in Fig. 4.2 and what was provided by Shashank (2016) in his report, the automotive industry takes some time to start using a new material. For years, original equipment manufacturers (OEMs) have been using certain materials in the automotive industry. If these materials were to change, this will require some changes to be made in the manufacturing processes. These changes might be expensive and OEMs might not see the sustainable reasons behind shifting from conventional glass fibers to natural fiber in composites; hence, making it hard to convince them. Moreover, the ability to reliably predict crash-worthiness, and durability of a material is not always easy and it is reflected on the product cost. Also, it should be taken into consideration that the automotive industry is very sensitive to cost and low profit margins.

It should be taken into consideration that the availability of new material is one of the important things that affect its usage. If the material was tested and had the desired properties but only available at one supplier, it will not be considered. A single material source will make it hard to always have a large quantity readily

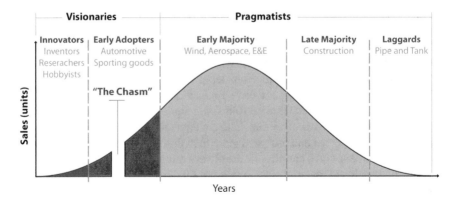

Fig. 4.2 Adoption life cycle of natural fiber composites (Midani et al. 2019)

available (Shashank 2016). Unfortunately, natural fibers are not available worldwide, as discussed earlier, where only 17 countries are responsible for 95% of the annual world production of natural fibers as shown in Fig. 1.5 (FAO 2021). Furthermore, there is large variability in the properties of the fibers which is reflected on the composite properties. Therefore, these points should be considered when dealing with NFC.

Also what actually hinders the market growth of natural fiber composites in the automotive industry is their hydrophilicity and the weak interfacial adhesion between the fibers and the matrix (Suddell and Evans 2005; Carruthers and Quarshie 2014). The final properties of fibers are affected by several controlled and uncontrolled factors during their processing (Elseify et al. 2019). Additionally, natural fibers are thermally unstable (<230 °C) which limits their compatibility with thermoplastics matrices having high melting temperature such as PET and PA. Some of the attempts made to improve properties of natural fibers were made by Fraunhofer scientists in collaboration with their partners. A special treatment was used to make natural fibers more hydrophobic. This will solve one of natural fibers' problems, and natural fibers can be used more widely (TechnicalTextile.net 2018).

One of the drawbacks of natural fibers is that they degrade when exposed to different environmental conditions. The degradation modes could be in the form of mechanical, biological, thermal, hydro, or due to weather degradations. Biological degradation is in the form of oxidation, reduction, and hydrolysis. The degradation occurs as a result of an attack by biological organisms or enzymes. The biological organisms could attack hemicellulose in the fibers which results in their hydrolysis into digestible units. The enzymes could also cause the degradation of fibers through oxidation and reduction reactions (Rowell 1998; Kumar et al. 2019). The water absorption is mainly caused by the void content in the fibers and also due to the presence of non-crystalline parts. Hemicellulose is the chemical constituents that absorbs the largest amount of water/moisture; however, cellulose and lignin contribute to the water absorption. Water degradation takes the forms of swelling, shrinking, and cracking. When the fiber absorbs moisture it swells, then it shrinks back again when it dries. This swelling/shrinking behavior happens simultaneously within the composite causing uneven swelling, thus composite cracking. Moreover, the mechanical properties are affected when natural fibers absorb water, since the water molecules react with the OH groups in the fibers leading to fiber weakening (Rowell 1998; Dhakal et al. 2016; Kumar et al. 2019). Weather degradations are caused by the exposure of NFC to wind, dust, snow, or ultraviolet radiations. This type of degradation is mainly caused by lignin which is very sensitive to ultraviolet radiations (Kumar et al. 2019). As for thermal degradation, natural fibers are known to have low thermal stability and are highly flammable. Hence, the composites reinforced with natural fibers would have insufficient fire resistance. Cellulosic fibers in general undergo two main thermal degradation stages: one associated with hemicellulose and the other with lignin. Hemicellulose degradation stage is normally between 200 and 300 °C, and lignin degradation stage is between 300 and 400 °C (Bachtiar et al. 2019; Kumar et al. 2019). Consequently, thermal degradation leads to reduction in the mechanical properties as a result of stresses, cracks, fractures, or abrasion (Kumar et al. 2019).

Researchers must think of ways or techniques to minimize the variability of natural fiber properties. These research attempts could be in the form of developing a new weaving technique, for example, like the ones developed by Bcomp or by developing a manufacturing technique that is suitable for natural fibers. Additionally, researchers should think outside the box, meaning that they should understand that what is suitable for man-made fibers is not necessarily suitable for the processing of natural fibers.

4.2   Performance Requirements

For any material to be approved by the automotive industry, it has to first meet several criteria. Mostly, these criteria are obtained from governmental regulations and legislations that are concerned with environmental and safety issues. Moreover, the customer requirements play an important role in setting the performance requirement criteria. The most important criterion is lightweight. However, one of the main obstacles is the high cost of lightweight materials. Additionally, it should be noted that the cost of a new material is very important in deciding whether this material can be used in the automotive industry or not. The cost includes cost of raw material, manufacturing process, and product testing. The third most important criterion is the crashworthiness. Crashworthiness is the ability of the material to absorb energy with controlled failure and gradual decay when subjected to impact. Furthermore, stiffness, strain, elongation, and strain at break are also very important to understand the deformation behavior of the material (Ghassemieh 2011).

The most important automotive market specifications are (Suddell and Evans 2005; Bledzki et al. 2006):

- Lightweight
- Ultimate breaking force and elongation
- Flammability
- Flame retardancy
- Acoustic absorption
- Flexural properties
- Impact strength
- Odor
- Water absorption
- Dimensional stability
- Suitability for processing
- Crash behavior.

Natural fiber composites are environment-friendly lightweight materials which can be used safely in the automotive industry after some modification and tuning. They could also be coupled with flame retardants to enhance their thermal properties.

4.3 Manufactured Parts

The well-established automotive interior parts produced and already in use nowadays are door panels, front and rear door liners, headliners, rear parcel shelves, seat backs, spare tire covers, and other interior trims made of flax, hemp, and kenaf-reinforced polypropylene matrix with 50% weight fraction (Suddell and Evans 2005). Figure 4.3 shows a door panel produced by UFP Technologies, Inc. made from polypropylene reinforced with 50% natural fibers (UFP Technologies 2019).

Table 4.1 shows a summary of automotive parts manufactured from natural fiber composites classified by car manufacturer. Mercedes-Benz in recent years has invested a large amount of money in research to develop recyclable sustainable automotive parts (Suddell and Evans 2005; Akampumuza et al. 2017), and this can be seen in Table 4.1 where different parts were made with natural fibers as reinforcements.

However, the above-mentioned parts in Table 4.1 were not the only ones manufactured using natural fibers. In fact, there are other successful attempts by companies like Bcomp, Porsche Motorsport, and Polestar. As mentioned earlier in this section, Bcomp has invented a grid that provides the composite with high stiffness and minimum weight. In a recent article on the emergence of flax fibers, Bcomp powerRibs technology was illustrated.

In 2018, Bcomp powerRibs and ampliTex flax fabrics were used in the Tesla S P100D race car to manufacture body panels, shown in Fig. 4.4a. The use of natural fibers reduced the weight by 500 kg. SPV Racing was the 1st to buy the EGT Tesla model S P100D. They worked on developing the race car till it became the only FIA-approved Tesla car, shown in Fig. 4.4b (Bcomp 2020a).

Polestar is an electric performance car brand. They aim in reducing the amount of plastic used in their cars. They used Bcomp powerRibs and ampliTex in the manufacturing of interior panels, shown in Fig. 4.5. When natural fibers were used, the weight of the panels was reduced by 50% and the plastic content was reduced by 80% (Polestar 2020).

McLaren Racing has collaborated with Bcomp to develop the first natural fiber composite racing seat for the car F1. Using reverse engineering, Bcomp succeeded in replacing carbon fibers with flax fibers as shown in Fig. 4.6. As mentioned on

Fig. 4.3 Door panel produced by UFP Technologies, Inc. made with 50% natural fibers and polypropylene (UFP Technologies 2019)

Table 4.1 Automotive parts reinforced with natural fibers

Manufacturer	Part	Fiber type	Matrix	Reference
Ford	Front grill	Hemp	Polypropylene	Akampumuza et al. (2017)
	Sliding door inserts	Wood	–	Holbery and Houston (2006)
Mercedes-Benz	Door panels	Flax	–	Norhidayah et al. (2014)
	Door panels	Sisal	–	Norhidayah et al. (2014)
	Door panels	Jute	–	Akampumuza et al. (2017)
	Door panels	Flax/sisal	Epoxy	Akampumuza et al. (2017)
	Engine encapsulations	Flax	Polyester	Akampumuza et al. (2017)
	Parcel shelves	Flax	–	Akampumuza et al. (2017)
	Rear trunk covers	Flax	–	Akampumuza et al. (2017)
	Trim strip wood veneers	Flax	–	Akampumuza et al. (2017)
	Rear panel shelves	Sisal	–	Akampumuza et al. (2017)
	Spare tire wheel covers	Abaca	Polypropylene	Koronis et al. (2013), Akampumuza et al. (2017)
	Seat back rest	Coconut	–	Norhidayah et al. (2014)
	Instrument panel support	Wood	–	Norhidayah et al. (2014)
	Trunk panel	Cotton	–	Norhidayah et al. (2014)
	Under floor protection	Banana	–	Norhidayah et al. (2014)
	Floor panel	Flax	–	Norhidayah et al. (2014)
	Insulation	Cotton	–	Norhidayah et al. (2014)
Toyota	Spare tire cover	Kenaf	Polylactic acid	Akampumuza et al. (2017)
	Floor mats	Kenaf	–	Norhidayah et al. (2014)

(continued)

Table 4.1 (continued)

Manufacturer	Part	Fiber type	Matrix	Reference
	Spare tire cover	Kenaf	–	Norhidayah et al. (2014)
	Luggage compartment	Bamboo	–	Norhidayah et al. (2014)
	Lexus package shelves	Kenaf	–	Holbery and Houston (2006)
Volkswagen	Door panel	Flax	–	Norhidayah et al. (2014)
	Door panel	Sisal	–	Norhidayah et al. (2014)
	Door inserts	Natural fibers	–	Akampumuza et al. (2017)
	Seat backs	Natural fibers	–	Akampumuza et al. (2017)
	Rear flap lining	Natural fibers	–	Akampumuza et al. (2017)
	Parcel trays	Natural fibers	–	Akampumuza et al. (2017)
	Door trim panel	Flax/sisal	Polyurethane	Akampumuza et al. (2017)
General Motors	Acoustic insulation	Cotton	–	Norhidayah et al. (2014)
	Ceiling liner	Kenaf	–	Norhidayah et al. (2014)
	Door panel	Hemp, flax, kenaf	–	Holbery and Houston (2006), Norhidayah et al. (2014)
	Seat backs	Hemp, flax, kenaf	–	Norhidayah et al. (2014)
	Seat backs	Wood fibers	–	Akampumuza et al. (2017)
	Cargo area floor	Wood fibers	–	Akampumuza et al. (2017)
	Door panel	Flax	Polypropylene	Akampumuza et al. (2017)
BMW	Door panel	Sisal		Norhidayah et al. (2014)
	Sound proofing	Cotton		Akampumuza et al. (2017)
	Seat back rest	Wood fibers	–	Akampumuza et al. (2017)

(continued)

Table 4.1 (continued)

Manufacturer	Part	Fiber type	Matrix	Reference
	Door panel	NF prepreg	Acrodur	Nv (2014), Akampumuza et al. (2017)
Mitsubishi	Floor mats	Flax	Polylactic acid/nylon	Akampumuza et al. (2017)
	Indoor cladding	Hemp/cotton	–	Akampumuza et al. (2017)
	Seat back lining	Hemp/cotton	–	Akampumuza et al. (2017)
	Floor panel	Hemp/cotton	–	Akampumuza et al. (2017)
Mitsubishi + Fiat	Interior components	Bamboo	Polybutylene	Akampumuza et al. (2017)
Volvo	Dashboards	Hemp/jute	Rapeseed/soy resin	Akampumuza et al. (2017)
	Ceilings	Hemp/jute	Rapeseed/soy resin	Akampumuza et al. (2017)
Citroen	Parcel shelves	Recycled wood + vegetable fibers	–	Akampumuza et al. (2017)
	Boot linings	Recycled wood + vegetable fibers	–	Akampumuza et al. (2017)
	Door panels	Recycled wood + vegetable fibers	–	Akampumuza et al. (2017)
	Mud guards	Recycled wood + vegetable fibers	–	Akampumuza et al. (2017)
Honda	Floor area parts	Wood	–	Norhidayah et al. (2014)
Mazda	Door trims	Kenaf	Polypropylene	Akampumuza et al. (2017)
	Interior parts	Kenaf	Polypropylene	Akampumuza et al. (2017)

their Web sites, the car had no problems when tested in a pre-season testing. The flax fibers used offered the required strength while at the same time reducing weight and carbon footprint. Moreover, the addition of flax fibers has improved the vibration absorption and impact resistance 5 times better than flax fibers. Finally, the cost was reduced by nearly 30% when flax fibers were used instead of carbon fibers (McLaren 2020).

Fig. 4.4   a Naked Tesla race car S P100D showing a body made from natural fibers ©Bcomp and **b** SPV's Tesla approved race car (Bcomp 2020a) ©SPV Racing

Porsche Motorsport has won the JEC innovation award in 2019 for the car 718 Cayman GT4 Clubsport. The award was offered because they succeeded in replacing carbon fibers with Bcomp natural fibers in the manufacturing of door and rear wing. The door and rear wing in Porsche 718 Cayman GT4 CS were made using RTM and an autoclave process, respectively. They again collaborated with Bcomp to make a full natural fiber body kit presented in Porsche 718 Cayman GT4 CS MR, shown in Fig. 4.7. Both cars were reinforced with Bcomp powerRibs and ampliTex. The cars were manufactured using same manufacturing technologies and molds used with carbon fibers. Flax fibers gave the car the needed performance with minimal weight.

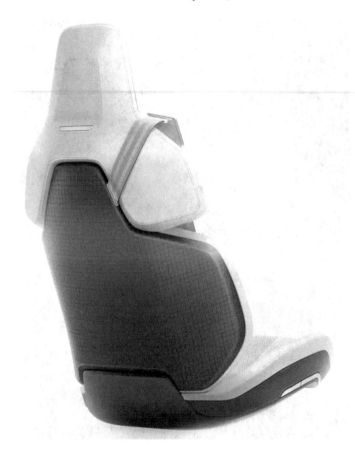

Fig. 4.5 Polestar precept interior panels reinforced with Bcomp flax fibers (Polestar 2020)

Natural fiber composites showed ductile crash behavior with no sharp shattering. Replacing carbon fibers with natural fibers reduced the cost by 30% and improved sustainability of the final parts by lowering carbon footprint by 75 (JEC Group 2019; Bcomp 2020b, c; Four Motors 2020; Porsche Motorsport 2020).

Generally, Motorsport industry is playing a very important role in facilitating the diffusion of natural fibers into the automotive industry. Once the car performance proved its reliability, the technology could be applied to high-volume markets and be used in automotive, aerospace, and marine applications.

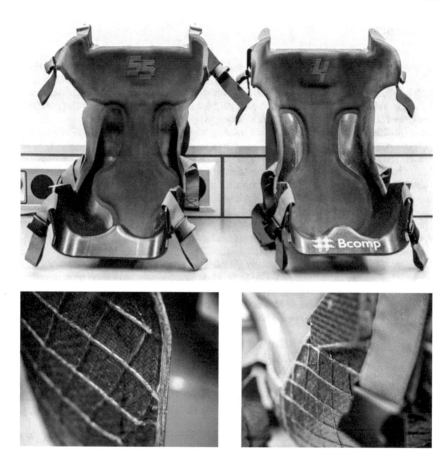

Fig. 4.6 McLaren F1 seat reinforced with Bcomp powerRibs (McLaren 2020)

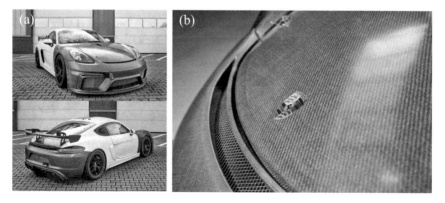

Fig. 4.7 Porsche Motorsport 718 Cayman GT4 CS MR with full natural fiber body kit © 2020 Dr. Ing. h.c. F. Porsche AG © Bcomp/Johannes Nollmeyer

References

Akampumuza O, Wambua PM, Ahmed A et al (2017) Review of the applications of biocomposites in the automotive industry. Polym Compos 38:2553–2569. https://doi.org/10.1002/pc.23847

Bachtiar EV, Kurkowiak K, Yan L et al (2019) Thermal stability, fire performance, and mechanical properties of natural fibre fabric-reinforced polymer composites with different fire retardants. Polymers (Basel) 11. https://doi.org/10.3390/polym11040699

Bcomp (2020a) Racing the EGT Tesla and building a sustainable racetrack. https://www.bcomp.ch/news/racing-egt-tesla-and-building-sustainable-racetrack/

Bcomp (2020b) Porsche uses sustainable alternative to carbon fibres for Nürburgring 24h race. https://www.bcomp.ch/?url=https%3A%2F%2Fwww.bcomp.ch%2Fnews%2Fporsche-full-natural-fibre-bodykit%2F. Accessed 3 Feb 2021

Bcomp (2020c) Porsche Motorsport innovates with series production of bio-based composites. https://www.bcomp.ch/news/porsche-innovates-with-series-production-of-bio-based-composites/. Accessed 3 Feb 2021

Bledzki AK, Faruk O, Sperber VE (2006) Cars from bio-fibres. Macromol Mater Eng 291:449–457. https://doi.org/10.1002/mame.200600113

Carruthers J, Quarshie R (2014) Technology overview biocomposites. Knowl Transf Netw 70

Dhakal HN, MacMullen J, Zhang ZY (2016) Moisture measurement and effects on properties of marine composites. Elsevier

Elseify LA, Midani M, Shihata LA, El-Mously H (2019) Review on cellulosic fibers extracted from date palms (Phoenix Dactylifera L.) and their applications. Cellulose 6. https://doi.org/10.1007/s10570-019-02259-6

FAO (2021) FAOSTAT. http://www.fao.org/faostat/en/#data/QC. Accessed 19 Feb 2021

Four_Motors (2020) FOUR MOTORS bioconcept-car. https://www.fourmotors.com/2020/09/24/four-motors-startet-mit-weltweit-erstem-vollnaturfaser-karosseriebausatz-ins-24h-rennen/. Accessed 3 Feb 2021

Ghassemieh E (2011) Materials in automotive application, state of the art and prospects. In: Chiaberge M (ed) New trends and developments in automotive industry. IntechOpen, UK, pp 365–394

Holbery J, Houston D (2006) Natural-fiber-reinforced polymer composites in automotive applications. JOM 58:80–86. https://doi.org/10.1007/s11837-006-0234-2

JEC Group (2019) Porsche innovates with series production of bio-based composites. http://www.jeccomposites.com/knowledge/international-composites-news/porsche-innovates-series-production-bio-based-composites

Koronis G, Silva A, Fontul M (2013) Green composites: a review of adequate materials for automotive applications. Compos Part B Eng 44:120–127. https://doi.org/10.1016/j.compositesb.2012.07.004

Kumar R, Ul Haq MI, Raina A, Anand A (2019) Industrial applications of natural fibre-reinforced polymer composites–challenges and opportunities. Int J Sustain Eng 12:212–220. https://doi.org/10.1080/19397038.2018.1538267

McLaren (2020) Revealed: how McLaren is pioneering the use of sustainable composites in F1. https://www.mclaren.com/racing/team/natural-fibre-sustainable-composite-racing-seat/. Accessed 21 Sep 2020

Midani M, Hassanin A, Hamouda T (2019) Where are natural fibre composites heading? JEC Compos

Norhidayah MH, Hambali AA, Yuhazri YM et al (2014) A review of current development in natural fiber composites in automotive applications. Appl Mech Mater 564:3–7. https://doi.org/10.4028/www.scientific.net/AMM.564.3

Nv L (2014) Flax/Acrodur® sandwich panel 73–78

Polestar (2020) Polestar and Bcomp. https://www.polestar.com/us/news/polestar-and-bcomp/. Accessed 3 Feb 2021

Porsche_Motorsport (2020) 718 Cayman GT4 clubsport. https://motorsports.porsche.com/internati onal/en/article/718caymangt4clubsport. Accessed 3 Feb 2021

Rowell RM (1998) Property enhanced natural fiber composite materials based on chemical modification. Sci Technol Polym Adv Mater 717–732. https://doi.org/10.1007/978-1-4899-0112-5_63

Shashank M (2016) Material qualification in the automotive industry

Suddell BC, Evans WJ (2005) Natural fiber composites in automotive applications. In: Mohanty AK, Misra M, Drzal LT (eds) Natural fibers, biopolymers, and biocomposites. Taylor & Francis, pp 231–259

TechnicalTextile.net (2018) Fraunhofer scientists stop natural fibres absorb water. Fraunhofer scientists in collaboration with their partners have,water very easily which limits their mechanical strength. https://www.technicaltextile.net/news/fraunhofer-scientists-stop-natural-fibres-absorb-water-240553.html. Accessed 21 Feb 2021

UFP Technologies (2019) Natural Fibers. https://www.ufpt.com/materials/natural-fibers.html. Accessed 3 Dec 2019

Chapter 5
Sustainability Assessment and Recycling of Natural Fiber Composites

Abstract Greenwashing and false sustainability claims are being used extensively to benefit from the rise of the eco-conscious consumer. However, it is not always necessary for the products made with natural resources to be environmentally friendly. A comprehensive sustainability assessment is required to validate any sustainability claim. This chapter is devoted to the discussion of the sustainability assessment and recycling of natural fiber composites (NFC). The standard methods used in the sustainability assessment are introduced including life cycle assessment (LCA) and carbon footprint. A comparison between the carbon footprint of different natural fibers is conducted, as well as a comparison between the carbon footprint of NFC and glass fiber composites. Advantages and disadvantages of natural fibers and glass fibers throughout their life cycle are discussed. Finally, the different methods and routes for recycling NFC are explained.

Keywords LCA · Carbon footprint · Sustainability assessment · Natural fiber composites · Recycling

5.1 Life Cycle Assessment and Carbon Footprint

Life cycle assessment (LCA) and carbon footprint are two standardized methods that are used to assess the environmental impact of a certain product through its life cycle. However, they are not the same. LCA is concerned with assessing multiple environmental impacts of a product such as global warming, ozone depletion, acidification of soil and water, and human toxicity. On the other hand, carbon footprint has a sole focus which is the climatic change due to greenhouse gas (GHG) emissions. Hence, carbon footprint can be considered a subset of a complete LCA. In carbon footprint, the different gas emissions, such as carbon, sulfur hexafluoride, and methane, are converted into CO_2 equivalents (de Beus et al. 2019; Liebsch 2019). Table 5.1 shows the environmental impact indicators and their unit of measurements and description. LCA is a crucial step toward sustainability and development of a certain product. It is very important to reduce the greenhouse gas emissions because, as can be seen in

Table 5.1 Environmental impact indicators (Liebsch 2019)

Impact indicator	Unit	Description
Global warming	kg CO_2-eq	Indicator of potential global warming due to emissions of greenhouse gases to air
Ozone depletion	kg CFC-11-eq	Indicator of emissions to air that causes the destruction of the stratospheric ozone layer
Acidification of soil and water	kg SO_2-eq	Indicator of the potential acidification of soils and water due to the release of gases such as nitrogen oxides and sulfur oxides
Eutrophication	kg PO_3^--eq	Indicator of the enrichment of the aquatic ecosystem with nutritional elements, due to the emission of nitrogen- or phosphor-containing compounds
Photochemical ozone creation	kg ethene-eq	Indicator of emissions of gases that affect the creation of photochemical ozone in the lower atmosphere (smog) catalyzed by sunlight
Depletion of abiotic resources – elements	kg Sb-eq	Indicator of the depletion of natural non-fossil resources
Depletion of abiotic resources – fossil fuels	MJ	Indicator of the depletion of natural fossil fuel resources
Human toxicity	1,4-DCB-eq	Impact on humans of toxic substances emitted to the environment (Dutch version of EN15804 only)
Freshwater aquatic ecotoxicity	1,4-DCB-eq	Impact on freshwater organisms of toxic substances emitted to the environment (Dutch version of EN15804 only)
Marine aquatic ecotoxicity	1,4-DCB-eq/sup>	Impact on seawater organisms of toxic substances emitted to the environment (Dutch version of EN15804 only)
Terrestrial ecotoxicity	1,4-DCB-eq	Impact on land organisms of toxic substances emitted to the environment (Dutch version of EN15804 only)
Water pollution	m^3	Indicator of the amount of water required to dilute toxic elements emitted into water or soil (French version of EN15804 only)

(continued)

Table 5.1 (continued)

Impact indicator	Unit	Description
Air pollution	m^3	Indicator of the amount of air required to dilute toxic elements emitted into air (French version of EN15804 only)

Fig. 5.1, greenhouse gases are responsible of capturing most of the sun radiations. Consequently, raising the temperature of the atmosphere (Ecochain 2020).

The product life cycle is formed from 5 steps as shown in Fig. 5.2; raw material, manufacturing and processing, transportation, usage, and finally waste disposal or recycling. However, the product life cycle can be observed from three points of view: cradle-to-gate, cradle-to-grave, and cradle-to-cradle. Cradle-to-gate assesses the product from the raw material stage till it reaches the consumer; before consumption. Cradle-to-grave assesses the product till its disposal stage. Finally, cradle-to-cradle is similar to the cradle-to-grave but the last step is replaced by recycling process where the product could be used again, hence closing the cycle. Carbon footprint is considered a cradle-to-gate assessment. LCA follows two standards; ISO 14044 and ISO 14040. While carbon footprint can be applied through three standards: PAS 2050, GHG Protocol, and ISO 14067 (de Beus et al. 2019; Liebsch 2019).

The carbon footprint of flax, hemp, jute, and kenaf was measured by a research team in nova institute in 2019. Table 5.2 shows the carbon footprint values, measured in CO_2-eq/tonne, based on mass allocation and economic allocation. In multi-output

Fig. 5.1 Greenhouse gas effect on earth (Ecochain 2020)

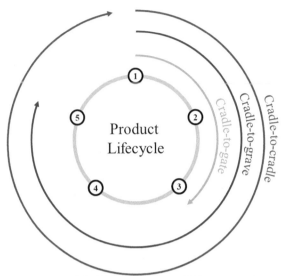

1: Raw material   2: Processing   3:Transportation
4: Retail & Use   5: Waste disposal / Recycling

Fig. 5.2 Product life cycle models

Table 5.2 Carbon footprint of flax, hemp, jute, and kenaf based on cradle-to-gate assessment (de Beus et al. 2019)

Fiber	Note	Carbon Footprint (CO_2-eq/tonne)	
		Mass allocation	Economic allocation
Flax	–	349	902
Hemp	Using mineral fertilizers	406	846
	Using organic fertilizers	364	759
	Using mineral + organic fertilizers	366	530
Jute	–	479	976
Kenaf	Traditional fiber production	418	975
	Based on DRÄXLMAIER Group	385	770

processes allocation occurs where the environmental burden is distributed over the different products. In case of natural fibers, allocation was needed since fibers are not the only product; the plant has other products like dust and shivers. Since the price of natural fibers is unstable and is dependent on several factors, mass allocation would be more stable than price-based allocation. After conducting the analysis, it was found that there was no significant difference between the results of the four fibers. The highest emissions were related to fertilizers. Using organic fertilizers have lower emissions than mineral fertilizers. However, organic fertilizers are not always applicable to all types of natural fibers. The values of carbon footprint based on the economic allocation are higher than those based on mass allocation since economic allocation increases the environmental impact of fibers over the other plant products. Additionally, the carbon uptake of the four natural fibers was found to be between 1.3 and 14 kg CO_2 per Kg of fibers. This amount of carbon dioxide will be released back at end of life of the product. Hence, the carbon footprint considering the carbon uptake was between -1100 and -600 CO_2-eq/tonne of natural fibers based on mass allocation. As for economic allocation, carbon footprint was found to be between -950 and zero CO_2-eq/tonne of natural fibers. Consequently, from cradle-to-gate natural fibers could be regarded as carbon sink. However, further processing of natural fibers results in an increase in the carbon footprint; nonetheless, natural fibers have lower carbon footprint than synthetic fibers (de Beus et al. 2019).

Comparing natural fibers to manmade fibers, it was found that natural fibers have 80% lower carbon footprint than glass fibers. However, natural fiber composites have 50% lower carbon footprint than glass fibers composites (de Beus et al. 2019). Moreover, natural fiber-reinforced composites consume 63% less energy than glass fiber-reinforced composites during the entire life cycle (Fitzgerald et al. 2021). Table 5.3 compares between the advantages and disadvantages of natural fibers and glass fibers through their life cycle.

As for the attempts made in the industry to reduce carbon footprint, in 2019 a life cycle documentation report was published on Mercedes-Benz website. Large quantities of coir fibers were used in the car body. They declared that the new B 180 car has lower carbon emissions than its predecessor as shown in Fig. 5.3. This is due to the lighter weight of B 180 car. When the entire life cycle was taken into consideration, the new B-class B 180 car showed 11% lower CO_2 emissions than its predecessor (Daimler 2019).

However, it is not always necessary for the products made with natural resources to be environmentally friendly. This is because cultivation and extraction of fibers have environmental impact and further processing of natural fibers increases the carbon dioxide emissions (Ita-Nagy et al. 2020). Therefore, it cannot be automatically assumed that any product from natural fibers will give a more environmentally benign product than manmade fibers just because natural fibers emit less carbon dioxide per Kg of fibers. However, a complete cradle-to-cradle LCA is needed to label a product sustainable (Fitzgerald et al. 2021). A study was made to compare the performance of flax–polyester composite to E glass–polyester composite in the manufacturing of wind turbine blades. It was found that for the flax composite to give the same performance of glass composite more fibers are required. This addition in material and processing will in return increase the environmental impact of flax composite (Shah et al. 2013; Fitzgerald et al. 2021).

Table 5.3 Advantages and disadvantages of natural fibers and glass fibers through their life cycle (Fitzgerald et al. 2021)

	Natural fibers	Glass fibers
Advantages		
Cradle	– High specific mechanical properties – Abundant – Renewable – Carbon absorbent – Low energy consumption	– High specific mechanical properties – Abundant – Low cost – Non-corrosive
Gate	– Low emission – Low energy consumption – Non-abrasive	– Well-established industry – Streamlined process
Use	– Extremely lightweight – Non-deleterious to health – Good insulators	– Durability – Lightweight – High operating temperatures
Grave	– Low emissions – Low energy consumption – Compostable – Biodegradable	None
Disadvantages		
Cradle	– Immature supply chain – Moisture absorption	– Non-renewable – Deleterious to health
Gate	– Moisture absorption – Incompatibility with matrices – Limited processing temperature – Abrasive	– Deleterious to health – High emissions – High energy consumption
Use	– Large properties variability – Moisture absorption – Limited processing temperature – Durability – Flame resistance	– Deleterious to health – Poor insulators
Grave	None	– Difficulty to recycle – Non-biodegradable

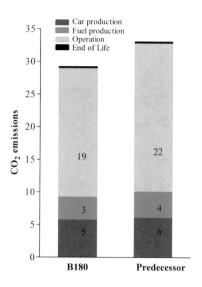

Fig. 5.3 Comparison between carbon dioxide emissions of Mercedes-Benz B-Class B 180 and its predecessor

5.2 Recycling of Natural Fiber Composites

The recycling of composites can be achieved mechanically, chemically, or biologically. Choosing the recycling route is dependent on the matrix type. Mechanical recycling depends on grinding, milling, or crushing the material. In case of thermoplastic matrix, the shredded material is subjected to melting using extrusion or injection molding process. The mechanical recycling process can be classified into two categories: primary and secondary mechanical recycling. The primary mechanical recycling route is usually used when the thermoplastic is not contaminated and does not require further purification before reuse. On the other hand, the secondary mechanical recycling route is adapted when the intended recycling material is a consumed product such as plastic bags and packages. The secondary mechanical process is more complex than the primary mechanical process. Antioxidants and stabilizing additives are used to compensate for the degradation due to the recycling process (Ramzy 2018; Cestari et al. 2018).

Chemical recycling, which is also known as tertiary recycling, can take several forms; hydrolysis, gasification, pyrolysis, and hydrocracking. The main aim of the process is to recover monomers through chemical reactions under different conditions to recover materials to be used again as raw materials for fuel and polymer industry. Hydrolysis process breaks down large molecules into smaller molecules in the presence of water. Gasification is the breakage of carbon-containing polymer molecules into carbon monoxide, hydrogen, and other gases in the presence of air or oxygen. Pyrolysis is carried out in oxygen-lean environment. Finally, hydrocracking is a two-stage process that uses high partial pressure to convert high boiling chemicals into low boiling ones (Ramzy 2018; Cestari et al. 2018).

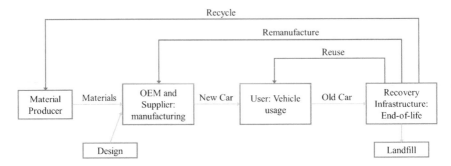

Fig. 5.4 Life cycle of a vehicle

Biological recycling is the biodegradation of polymers using microorganisms, air, and water. It is used to recycle biosynthesized polymers such as cellulose and polylactic acid. However, not all bioplastics are biodegradable and likewise not all synthetic polymers are non-biodegradable (Ramzy 2018).

The recycling processes used with polymeric materials are already well-established. However, the recycling of natural fiber-reinforced composites is still under research. Yet, there are guidelines and regulations set by the European Union and Asian countries that state that a significant percentage of the vehicles should be recycled or reused. Figure 5.4 shows the life cycle of a vehicle (Ghassemieh 2011). Hence, for any new material to get approval for use in the automotive industry its recycling options should be discussed and studied first.

References

Cestari SP, Silva Freitas D de F, Rodrigues DC, Mendes LC (2018) Recycling processes and issues in natural fiber-reinforced polymer composites. In: Koronis G, Silva A (eds) Green composites for automotive applications. Woodhead Publishing, pp 285–299

Daimler A (2019) 360° Environmental check Mercedes-Benz B-Class. Stuttgart, Germany

de Beus N, Carus M, Barth M (2019) Carbon footprint and sustainability of different natural fibre for biocomposites and insulation material. Nov Inst 57

Ecochain (2020) Ecochain. https://ecochain.com/. Accessed 17 Feb 2021

Fitzgerald A, Proud W, Kandemir A et al (2021) A life cycle engineering perspective on biocomposites as a solution for a sustainable recovery. Sustain 13:1–25. https://doi.org/10.3390/su1303 1160

Ghassemieh E (2011) Materials in automotive application, state of the art and prospects. In: Chiaberge M (ed) New trends and developments in automotive industry. IntechOpen, UK, pp 365–394

Ita-Nagy D, Vázquez-Rowe I, Kahhat R et al (2020) Life cycle assessment of bagasse fiber reinforced biocomposites. Sci Total Environ 720. https://doi.org/10.1016/j.scitotenv.2020.137586

Liebsch T (2019) Life cycle assessment (LCA)—Complete beginner's guide. In: Ecochain. https://ecochain.com/knowledge/life-cycle-assessment-lca-guide/. Accessed 10 Apr 2021

Ramzy A (2018) Recycling aspects of natural fiber reinforced polypropylene composites. Doctoral thesis, pp 1–146

Shah DU, Schubel PJ, Clifford MJ (2013) Can flax replace E-glass in structural composites? A small wind turbine blade case study. Compos Part B Eng 52:172–181. https://doi.org/10.1016/j.compositesb.2013.04.027

Chapter 6
Future Trends in Natural Fiber Composites in the Automotive Industry

Abstract Research on natural fiber composites (NFCs) has identified poor compatibility between natural fiber reinforcement and synthetic matrix, as one of the major factors limiting the wide use of NFC. A novel composite is introduced to resolve this problem, which is made entirely of cellulose and known as all-cellulose composite (ACC). ACC is believed to be the future of sustainable composites in the automotive industry. This chapter is devoted to the discussion of all-cellulose composites and its potential use as a sustainable composite in the automotive industry of the future. The manufacturing of ACC using conventional impregnation method (CIM) and partial dissolution method (PDM) is described, including a comparison between both methods. Then, the dissolution and regeneration stages and the types of solvents and anti-solvents used are also presented. Finally, the effect of the processing parameters on the properties of ACC is detailed and discussed.

Keywords All-cellulose composites · Conventional impregnation method · Partial dissolution method · Automotive

6.1 All-Cellulose Composites

One of the most sustainable forms of composites is the all-cellulose composite (ACC), and it is believed that ACC will have a very promising future. ACC is a composite made entirely form cellulose; the matrix and reinforcements are both obtained from cellulose-rich materials. ACCs are similar to all-polymer composites. Cellulose exists in four polymorphs: cellulose I, cellulose II, cellulose III, and cellulose IV. Cellulose I, which is referred to as native cellulose, exists in nature and could be extracted from plants and natural resources. Cellulose II is formed by the mercerization or regeneration of cellulose I. By the treatment of cellulose I and cellulose II using diamine treatment, cellulose III could be obtained. Finally, cellulose IV is formed by the heat treatment of cellulose III in glycerol. It should be noted that cellulose I has the highest crystallinity among the four polymorphs. In the literature, ACCs were made using either cellulose I or cellulose II. However, this research focuses on natural fibers; hence, the work is mainly concerned with ACCs made from cellulose I. The

L. A. Elseify et al., *Manufacturing Automotive Components from Sustainable Natural Fiber Composites*, SpringerBriefs in Materials, https://doi.org/10.1007/978-3-030-83025-0_6

source of cellulose used manipulates the type of the produced ACCs. Isotropic ACCs are made from micro- and nano-fibrillated cellulose, while non-isotropic ACCs are made from biological or natural fibers (Baghaei and Skrifvars 2020).

The concept or the idea of ACCs was first introduced by Nishino et al. in 2004 (Nishino et al. 2004; Nishino and Peijs 2014). And the main aim of ACC is to solve the problem of the poor fiber–matrix interfacial bond that exists between polymers and natural fibers. In ACCs, the matrix and reinforcement are chemically compatible since both are made of cellulose. The incompatibility between matrices and reinforcements results in poor stress transfer which deteriorate the mechanical properties of the composite (Huber et al. 2012; Baghaei and Skrifvars 2020).

6.2 Manufacturing of ACC

ACC can be manufactured using two routes as shown in Fig. 6.1. The first route is a two-step method, which is called conventional impregnation method (CIM), where dissolved cellulose solvent is regenerated in the presence of undissolved cellulose. The second route is a one-step method, which is called selective dissolving method or partial dissolution method (PDM); it involves the partial dissolution of cellulose

(a) Impregnation Method

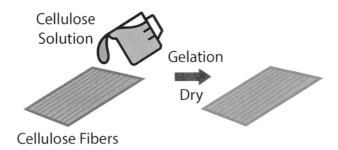

(b) Selective Dissolving Method

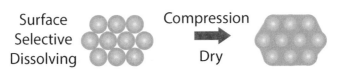

Fig. 6.1 All-cellulose composite using **a** impregnation method and **b** selective dissolving method

fiber surface then regenerated again in situ to act as a matrix and binds the fibers together. Comparing between the 2 methods, it was apparent that the PDM is more realistic to be used on an industrial scale than the CIM (Huber et al. 2012; Nishino and Peijs 2014).

However, everything comes with its advantages and disadvantages. The advantages of ACCs, as mentioned above, are the seamless chemical bond between the fibers and matrix, high mechanical properties, the good thermal behavior, non-toxicity, biocompatibility, biodegradability, sustainability, and environment-friendly. Moreover, ACCs have optical clarity and barrier properties better than other conventional properties. However, ACCs have some disadvantages. The processing of ACC is somehow difficult since cellulose has high viscosity even when used with low concentrations. Also, the ACCs made in the literature were all tried on very thin composites, less than 1 mm thin. Manufacturing a thick ACC is still under investigation and needs further research. Moreover, the cellulose source used should have high cellulose purity. Lignin and hemicellulose both can dissolve too with cellulose, leaving some impurities in the final composite. The impurities would act as inclusions, thus reducing the mechanical properties. Therefore, the removal of lignin and hemicellulose is a step that cannot be ignored. One of the challenging issues that faces ACCs is its hydrophilicity. Cellulosic fibers naturally are hydrophilic, which will make the composite prone to large quantities of water absorption (Huber et al. 2012; Nishino and Peijs 2014; Baghaei and Skrifvars 2020).

In both manufacturing routes, cellulose is dissolved using solvents and regenerated again using anti-solvents. Some of the solvents used are N-methylmorpholine-N-oxide (NMMO), lithium chloride/N,N-dimethylacetamide (LiCl/DMAc), ionic liquids (IL), and sodium hydroxide aqueous solutions. The most commonly used solvents are LiCl/DMAc and ILs. However, the use of LiCl/DMAc is limited to laboratory research due to some environment-related issues. On the other hand, ILs are regarded as green solvents and attracted attention due to their easy recyclability, negligible vapor pressure, and thermal stability and the ability to dissolve wide range of cellulose without needing any treatment or activation. Therefore, choosing the right solvent is an essential step to avoid using toxic, non-recyclable, or slow-dissolution one. As for anti-solvents, water, ethanol, methanol, acetone, or acetonitrile could be used. The chosen anti-solvent plays a role in the final properties of the composite. This step is considered to be one of the most important steps in the formation of ACC where hydrogen bonds are reformed between hydroxyl groups which result in cellulose regeneration. A study was made by Tan et al. to compare between water and ethanol as anti-solvents. It was found that using water was much better than ethanol since water allows for the formation of crystalline-oriented cellulose structure when used for regeneration (Baghaei and Skrifvars 2020). Additionally, the process of ACCs has some process parameters that needs to be optimized. The two most important parameters are the solvent immersion time and the cellulose regeneration rate or cellulose precipitation rate (Baghaei and Skrifvars 2020).

6.3 Effect of Processing Parameters

In 2008, Soykeabkaew et al. (2008) made ACC using PDM from ramie fibers. The fibers were aligned in 50 × 50 mm metal mold. The composite was prepared by immersing the fibers in distilled water at room temperature for 2 h. Then, the fibers were immersed in acetone for 2 h. Afterward, the fibers were immersed in 8 wt%/v LiCl/DMAc for 1, 2, 3, 4, 5, 6, or 9 h. After that, the fibers were removed from the solvent and exposed to ambient conditions for 12 h. Methanol as anti-solvent was used to remove the traces of LiCl and DMAc. The fibers were left in methanol for 12 h. Methanol was replaced by fresh one after 1 h and 6 h of immersion. Finally, the sample was dried into 2 steps: first, drying at room temperature for 12 h, and second, drying under vacuum at 60 °C for 24 h. The test results showed that increasing the immersion time in LiCl/DMAc provided better adhesion between fibers and matrix as shown in Fig. 6.2a. Moreover, increasing the immersion duration increased the optical clarity as shown in Fig. 6.2b. As for the mechanical properties of the fabricated ACCs, it was found that as the immersion time was increased, the samples became more brittle as indicated by Fig. 6.2c, which shown the fracture modes. However, after analyzing the stress–strain curves of the tested samples, it was noticed that increasing the immersion time reduced the tensile strength and the modulus of elasticity. On the other hand, increasing the immersion time of samples led to increasing their elongation. The thickness of the manufactured composites was not mentioned.

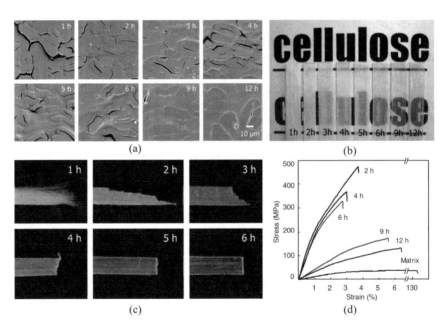

Fig. 6.2 a SEM of ACCs cross section, photographs shown, **b** optical clarity, **c** failed specimens, and **d** stress–strain curves of samples at different durations (Soykeabkaew et al. 2008)

In 2020, Chen et al. (2020) attempted the fabrication of ACC from flax fibers using PDM. Ionic liquid [EMIM] [OAs] was used as solvent. The authors specifically chose ILs over other solvents since it has the ability to dissolve cellulose at room temperature which is feasible and economical when applied on industrial scale. The aim of this work was to determine the effect of the IL on flax fibers to obtain the best process parameters. Unidirectional flax roving with low twist degree was used. The unidirectional flax yarn was aligned in 2 layers consisting of 145 yarns equivalent to 3 gm weight. The fibers were fixed at the ends to avoid shrinkage and fiber distortion. The fibers were impregnated with the solvent [EMIM] [OAs] under ambient conditions for different durations: 10, 25, 40, and 50 min. Afterward, the sample was pressed without applying temperature under 80 bars for 5 min. Then, specimens were washed with deionized water for 72 h to remove the ionic liquid solvent from fibers. The specimen drying took place in 2 steps; first, the samples were pressed at 80 °C and 50 bars for 2.5 h and then compressed for 0.5 h under 60 °C and 50 bars. Figure 6.3 shows the ACCs manufacturing steps. The results showed that increasing time and temperature of the IL accelerated the rate of cellulose dissolution. The dissolution rate was calculated using Arrhenius type law, similar to the one used with chemical reactions. The dissolution rate should be optimized such that only the primary and S1 layer of fibers are dissolved. Maintaining an undissolved S2 layer preserves the mechanical rigidity of fibers. In case of flax fibers, the S1 layer represents 10% of the fiber's thickness and could be dissolved in 20 min at room temperature. The XRD results showed cellulose II peak at 2θ = 12.2° for samples impregnated for 45 and 60 min. Moreover, there was no big difference between the crystallinity of the samples impregnated for 15 and 30 min. The crystallinity was found to decrease when the impregnation time was increased. The sample impregnated for 60 min had the lowest crystallinity indicating that the impregnation time was larger than needed which caused the dissolution of the S2 fiber

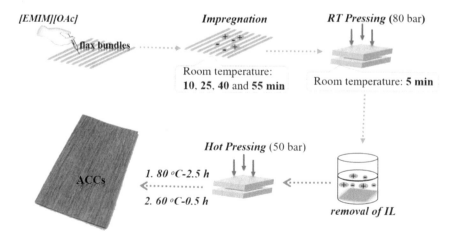

Fig. 6.3 Manufacturing steps of ACC (Chen et al. 2020)

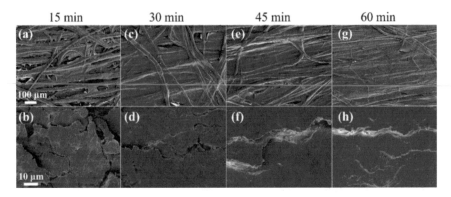

Fig. 6.4 SEM of (**a, c, e**, and **g**) fabricated ACCs and (**b, d, f**, and **h**) their cross sections at different immersion durations (Chen et al. 2020)

layer. These results were confirmed by the mechanical properties of composites. The samples impregnated for 60 min had the lowest mechanical properties compared to the other durations. The strength values obtained in this work, specifically the sample immersed for 45 min, were better than the literature with strength reaching up to 160 MPa and modulus up to 10 GPa. Additionally, it was found that increasing the immersion time increased the density of the produced composites. Moreover, increasing the immersion time improved the interfacial bond between fibers and matrix as shown in Fig. 6.4. Finally, it can be concluded that understanding the dissolution kinetics of cellulose fibers is essential because the aim is to create an ACC with high crystallinity and strength but with minimal conversion of cellulose I to cellulose II or amorphous cellulose. Hence, the best composite properties will be obtained when the solvent succeeded in dissolving the S1 layer of fibers without touching the S2 layer.

A completely biodegradable compatible composite can be made entirely from cellulose. ACCs have excellent properties compared to the other types of composites, and they represent the future of natural fiber composites in the automotive industry. ACC will solve the incompatibility of natural fibers and will provide a completely environmentally friendly auto part. It should be mentioned that the polymer matrix used in the composites has the highest environmental impact in the life cycle assessment (LCA) of natural fiber composites and not the fibers (Ramesh et al. 2020). Hence, ACC cellulose composite will provide a solution to this problem.

References

Baghaei B, Skrifvars M (2020) All-cellulose composites : a review of recent studies on structure, properties and applications

Chen F, Sawada D, Hummel M et al (2020) Unidirectional all-cellulose composites from flax via controlled impregnation with ionic liquid. Polymers (basel) 12:6–8. https://doi.org/10.3390/POLYM12051010

Huber T, Müssig J, Curnow O et al (2012) A critical review of all-cellulose composites. J Mater Sci 47:1171–1186. https://doi.org/10.1007/s10853-011-5774-3

Nishino T, Matsuda I, Hirao K (2004) All-cellulose composite. Macromolecules 37:7683–7687. https://doi.org/10.1021/ma049300h

Nishino T, Peijs T (2014) All-cellulose composites. In: Handbook of green materials. pp 201–216

Ramesh M, Deepa C, Kumar LR et al (2020) Life-cycle and environmental impact assessments on processing of plant fibres and its bio-composites: a critical review. J Ind Text 1–25. https://doi.org/10.1177/1528083720924730

Soykeabkaew N, Arimoto N, Nishino T, Peijs T (2008) All-cellulose composites by surface selective dissolution of aligned ligno-cellulosic fibres. Compos Sci Technol 68:2201–2207. https://doi.org/10.1016/j.compscitech.2008.03.023

Printed in the United States
by Baker & Taylor Publisher Services